Pollinators &
Native Plants
for Kids

An Introduction to Botany

Edited by Emily Beaumont and Brett Ortler

Cover and book design by Jonathan Norberg

Proofread by Andrew Mollenkof

Credits continued on page 143

Pollinators & Native Plants for Kids: An Introduction to Botany

Pollinators & Native Plants
for Kids

An Introduction to Botany

Jaret C. Daniels

Adventure PUBLICATIONS

an imprint of Adventure**KEEN**

Table of Contents

Mosses

Ferns

Trees

Shrubs

Grasses

Herbs

Green algae

Cycads

Liverworts

Introduction

Plants are amazing. They come in many different forms and can be found almost everywhere on Earth. As a group, plants are one of the most common forms of life on the planet. Many, like mosses, ferns, trees, shrubs, grasses, and herbs, are familiar organisms that we see most every day. Others, such as green algae, cycads, and liverworts, are a bit more unusual. In fact, the kinds of plants that are found in an area help define its habitat type. Examples include grasslands, forests, deserts, and savannas.

KNOW YOUR TAXONOMIC NAMES

Because so many lifeforms exist on Earth, scientists use taxonomy *(say it, tax-on-oh-mee)* to put living things in categories. **Taxonomy** is the science of naming and classifying organisms. It uses a ranked system that's a bit like a pyramid. At the top of the pyramid, there are big categories called kingdoms. You've probably heard of the animal kingdom, and plants belong to their own kingdom: Plantae *(say it, plan-tay)*, the plant kingdom. As you go down the pyramid, the levels get more specific, and there are fewer lifeforms in each category. Kingdom is followed by phylum, class, order, family, genus, and species.

In this system, called binomial nomenclature *(say it, bye-gnome-ee-ul gnome-en-clay-ture)*, each organism on Earth has what's called a scientific name. It has two parts. The first part of the scientific name refers to the **genus** and is a bit like a last name. It indicates which organisms are closely related, like your other immediate family members and you. The second part is called the specific epithet *(say it, ep-i-thet)*. It's like a first name, and it separates you from close relatives and identifies you as a unique individual or, in this case, as a unique **species.** The resulting taxonomic name only belongs to that lifeform. Scientific names are easy to spot because they are always written in *italics*. The whole system may sound complicated, but it isn't too hard to learn. In this book, you will see common names, such as purple coneflower, followed by scientific names, such as *Echinacea purpurea*.

Although many people regularly use common names, common names can be confusing. There can be more than one for the same organism, and people may use different names depending on where they live. This can make it hard to know if everyone is talking about the same thing. For example, a common southeastern wildflower, *Monarda punctata*, has many common names: dotted horsemint, dotted horse mint, spotted horsemint, or spotted beebalm. On the other hand, scientific names are known everywhere and standardized across regions, cultures, and languages. They are the global standard. It's also important to note that some

Monarda punctata, spotted beebalm

scientific names can change as scientists learn more about how organisms are related. Now, let's go back to *Monarda punctata* and look at its full taxonomy.

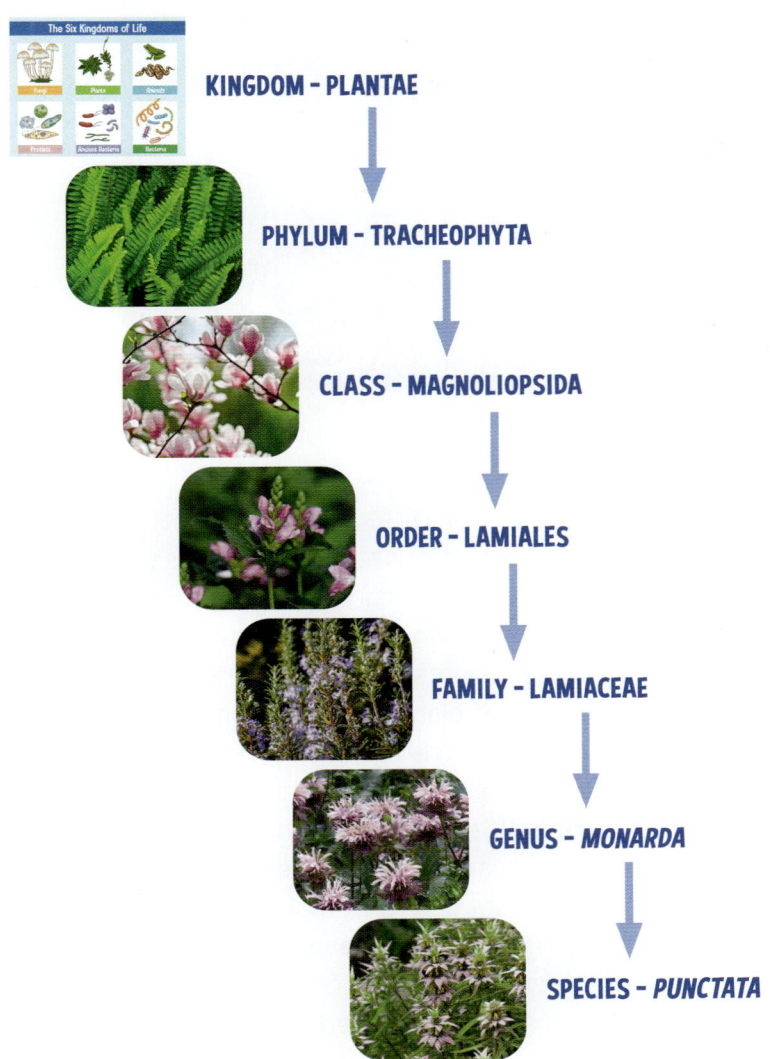

KINGDOM – PLANTAE

PHYLUM – TRACHEOPHYTA

CLASS – MAGNOLIOPSIDA

ORDER – LAMIALES

FAMILY – LAMIACEAE

GENUS – *MONARDA*

SPECIES – *PUNCTATA*

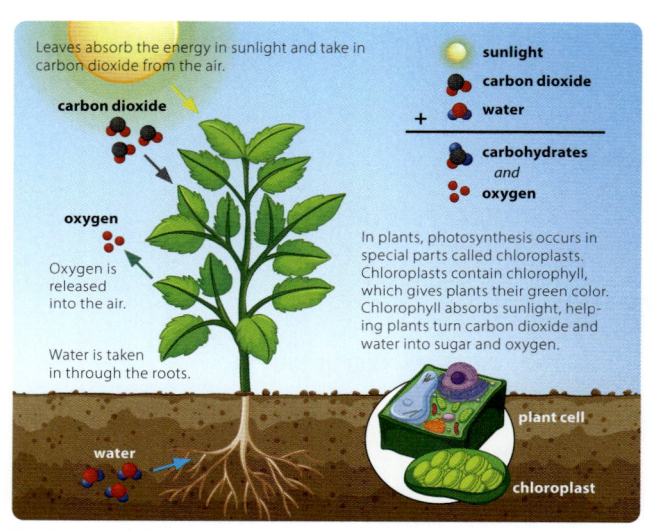

The basics of photosynthesis

HOW ARE PLANTS UNIQUE?

First of all, plants look different than animals, and they vary in several important ways: Plants produce their own food, and they can't move from one place to another on their own. Plants also have cell walls, and most possess roots, stems, and leaves. Another common feature of plants is their overall green color, which may be one of their most important features because it is critical to life on Earth. The color comes from **chlorophyll,** a green pigment found within **chloroplasts.** These are specialized organelles (structures) in plant cells that make photosynthesis possible. During **photosynthesis,** light energy from the sun, water taken up by roots, and carbon dioxide from the atmosphere are used to fuel a chemical reaction that produces sugar (glucose) and oxygen. Plants use the sugar as food to grow, then store any extra. This energy-rich resource also provides food for other organisms, like animals that are unable to produce their own food. The resulting oxygen that plants release into the atmosphere is essential for life on Earth, as most living things, including people, need oxygen to survive.

A BIT ABOUT BOTANY

Botany is the study of plants. It is a large scientific field that's linked to many other important science fields, including biology, medicine, and conservation. People who study plants are called **botanists,** and they work both in laboratories and out in nature. They study how plants help the Earth, identify new species, and discover how plants are related to each other. Botanists find safe ways to keep growing food; identify, understand, and fight plant pests and diseases; study fossils to understand past environments; and many other very useful things. They even help conserve rare and endangered species.

But you don't have to have a college degree to enjoy or study plants. Anyone can learn about plants and even collect information or make new discoveries. These people are called community (or citizen) scientists. Today, there are many ways for community scientists to help collect data and contribute to the study of plants. No matter if you live in the city or far out in the country, there are many community science programs available, and kids can participate too! Community scientists can help track invasive species, identify new areas where plants are growing, and even help understand how our changing climate affects plants.

THE FIRST PLANTS

Plants have been around on Earth for a very long time. In fact, they first appeared hundreds of millions of years before the earliest dinosaurs roamed. The

A green algae species under a microscope

ancient ancestors of all land plants evolved from multicellular algae (simpler plants consisting of many cells) in freshwater environments around 1 billion years ago. Evidence suggests that plants then moved onto land around 500 million years ago. These early plants were able to adjust their water content in a small way. They soon evolved a complicated network of tissues (vascular system) that transports water, minerals, and nutrients; this system also helps plants stay upright. Plants were then able to grow larger, make and store food, and live in a wide range of new and different habitats, leading to the evolution of complex habitats and ecosystems.

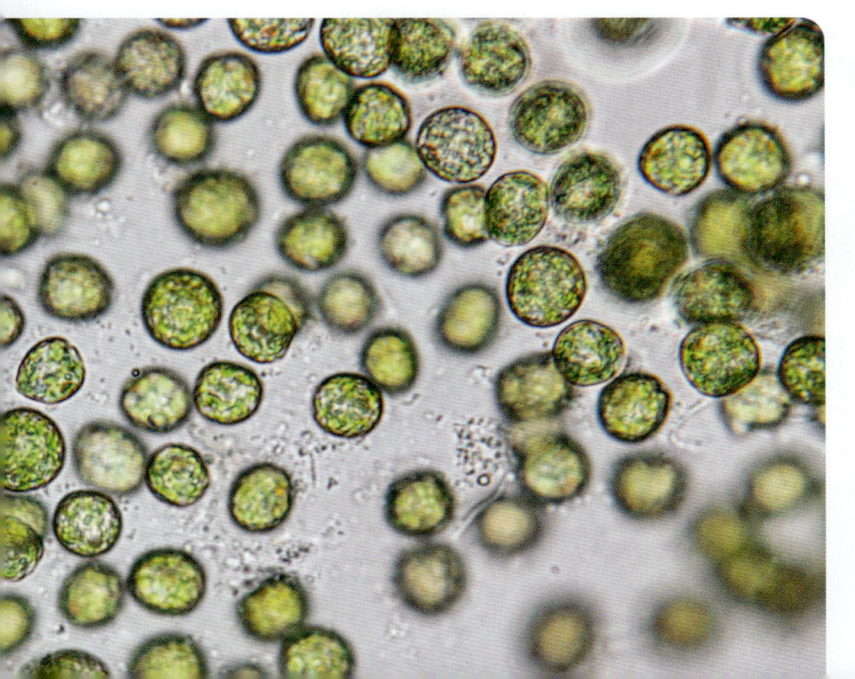

PLANT EVOLUTIONARY TIMELINE

Plants develop a protective outer cuticle, enabling them to colonize land ~500 million years ago

Plant vascular system evolved, enabling transport of water and nutrients ~430 million years ago

PRECAMBRIAN ERA

PALEOZOIC ERA

Freshwater algae and algae-like plants, possibly > 1 billion years ago

Nonvascular plants, including mosses and liverworts, appeared ~470 million years ago

Ferns, horsetails, and their allies appeared; early woody plants ~400 million years ago

PRECAMBRIAN ERA: 4.6 billion to 541 million years ago
PALEOZOIC ERA: 541 million to 252 million years ago
MESOZOIC ERA: 252 million to 66 million years ago

Earliest seed plants, gymnosperms appeared ~380 million years ago

First dinosaurs appeared ~240 million years ago

MESOZOIC ERA

First flowering plants appeared ~135 million years ago (possibly much earlier)

Cycads, conifers appeared; ginkgo soon thereafter ~300 million years ago

GYMNOSPERMS VS. ANGIOSPERMS

Gymnosperms *(say it, jim-no-sperms)* and angio-sperms *(say it, angie-o-sperms)* are the two major groups of seed-producing land plants on Earth. Gymnosperms are an ancient group. They first evolved more than 300 million years ago. Today, there are a total of just over 1,100 species of them; living examples include cycads, conifers, and Ginkgo biloba, a single, unique-looking tree species that is sometimes called a living fossil because it has barely changed in over 200 million years. Gymnosperms were the main plants on land during the age of the dinosaurs. The word **gymnosperm** comes from two Greek words that together mean "naked seed."

Gymnosperms do not produce flowers. Their seeds usually form as cones. For example, pine trees are gymno-sperms. The seeds are not enclosed in a fruit but grow on the surface of scales or leaves. The wind transports their pollen and helps with repro-duction. Most (but not all) living gymnosperms are also evergreen—they don't lose their leaves and stay green year-round.

White pines are a famous type of conifer.

Conifers—which include pine, fir, spruce, and cedar, to name a few—are the most diverse group of gymnosperms. They are often the dominant woody tree species in temperate and boreal forests. By far, the most obvious feature of conifers is their leaves. They tend to be small, quite simple in overall shape, and formed as needles or scales.

Angiosperms are the other major group of land plants that produce seeds. Also called flowering plants, they evolved much later into what most people picture when they think of typical plants today. The term angiosperm comes from two Greek words that mean "vessel seed" and refers to their seeds being enclosed by a fruit. Worldwide, there are over 300,000 flowering plant species—that's around 90% of all

Roses are a famous example of a flowering plant.

plant species on Earth. Flowering plants transformed terrestrial (land) systems and caused a tremendous explosion of life on Earth—and plants continue to support much of the planet's amazing number of different species today (biodiversity). In fact, flowering plants are a main factor in how we live, how we make money, and how we feed ourselves.

Flowers are the most obvious and showiest part of angiosperms. They have adapted purely for reproduction. Over time, they have continued to evolve into an amazing assortment of sizes, colors, and forms. Together with rich food rewards such as pollen, nectar, or fatty oil, they help attract animal pollinators. This diversification (variety) is believed to have fueled a similar diversification of insects. For example, today some types of plants rely on only one type of insect for pollination. Angiosperms dominate the planet on which we live.

THE ANATOMY OF A FLOWERING PLANT

Flowers serve a special role in plant survival. Without flowers, most plants could not reproduce (make more of themselves). Although flowers come in many forms, colors, and sizes, they all generally share a similar overall structure. Most flowering plants have what are called perfect flowers. They contain both male and female reproductive parts.

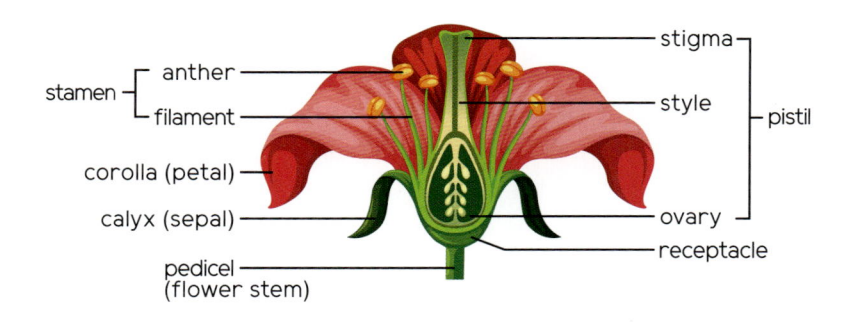

The showiest parts of a flower are typically the **petals.** All of the petals together make up the corolla. Below the petals are **sepals.** They represent the outermost parts of a flower and help protect the flower bud and eventually the developing fruit. All of the sepals together are called the **calyx.** While usually green in color and resembling leaves, they can sometimes be quite colorful. Many plants also have **bracts** below the calyx. These are modified leaf-like structures that can serve to protect flowers, fruits, or seeds from herbivores (plant eaters) or cold temperatures. In some species, they can also be enlarged or colorful to serve much the same role as petals for attracting pollinators. Poinsettias are a good example. Their large, reddish, petal-looking structures are actually modified bracts. They surround the actual flowers, which are tiny, yellowish in color, and located in the center.

If you look closely at the interior portion of a flower, you can spot the **stamen.** This represents the male reproductive part of a flower. It consists of a long stalk-like **filament** and a pollen-producing structure called an **anther** at the top. **Pollen** is typically a fine, grainy substance that often looks like powder. It contains the male reproductive cells and is often transported by wind, water, or an insect or bird. Pollen from some plants, such as ragweed, can also trigger seasonal allergies.

The female part of the flower is called the **pistil.** It is found in the center of the flower. It consists of a large base called the **ovary.** It supports a long **style,** which in turn is topped by a somewhat flat or bulb-like part called a **stigma.** The stigma has a sticky or waxy surface that is designed to receive pollen, helping with fertilization. Once fertilized, the ovary develops into a **fruit** containing one or more **seeds.**

Supporting the flower is an individual stalk called a **pedicel** *(say it, ped-ih-cell)*. It is attached to the **receptacle,** a large section or foundation of the pedicel to which all the flower parts are attached.

Flowers can be produced individually (single) or in clusters (groups). The arrangement of flowers on a stem is called an **inflorescence** *(say it, in-floor-ess-ence)*. Each flowering plant species makes a certain type of inflorescence. There are many different types, and their traits can help tell one species from another.

An oak, an example of a monecious tree

Not all flowering plants make perfect flow-ers. Some can be **monoecious** *(say it,*

A ginkgo tree, an example of a dioecious tree

mon-EE-shuss). They produce separate male and female flowers on the same plant. Many monoecious plants are pollinated by the wind. Common examples include corn, as well as oak, hickory, and pine trees. Plants that are monoecious have an advantage: they can self-pollinate, or successfully reproduce without the presence of another plant.

Other plants are either all-male or all-female. These are called **dioecious** *(say it, die-EE-shuss)*. Those plants need both a male and female plant to reproduce and make fruit. Examples include persimmon, holly, ginkgo, and kiwi. The main advantage dioecious plants have is that they can reproduce with unrelated plants, increasing their genetic diversity. That helps them adapt more quickly (survive in different environments).

PLANT LIFE CYCLES

Plants are typically classified as an annual, biennial, or perennial based on how long they take to complete their life cycle. Annual plants complete their entire life

cycle from seed to flower to seed in one growing season, after which the entire plant dies. Biennial plants take two years to complete their life cycle. They produce roots and vegetation in the first growing season, followed by flowers, fruit, and seeds in year two. Lastly, perennial plants live for more than two years and often much longer depending on the species. Perennials are typically divided into herbaceous plants and woody plants.

Large-leaved lupine seed pods

Herbaceous perennials, such as wildflowers, do not have persistent woody stems. The top vegetative portion of the plant typically dies back to the ground in winter, especially in colder regions, and new stems grow back the following year. Woody perennials, such as trees and shrubs, have persistent stems that do not die back each year. They may drop their leaves each year before winter, but the stems and branches remain year-round.

KNOW YOUR SOIL TYPES

Soil is the loose material on the upper surface of the Earth in which plants grow. It is a mix of minerals, rock, and organic material. Soil provides plants with nutrients, water, and structural support for their roots. Soil varies greatly in its physical and chemical makeup, as well as the type and arrangement of particles found in its structure. Soil structure is important for plant health and growth as it affects how water and air move through the soil. This, in turn, affects nutrient availability, root growth, and water retention.

In general, most soils can be categorized into four types. Sandy soil is light in weight, has a grainy texture, and is low in nutrients. Water drains quickly through sandy soil. Silty soil feels smooth and somewhat slippery when wet;

Knowing your soil type is essential when planning your garden.

it is high in nutrients and retains water longer than sand. It is nonetheless considered a well-drained soil. Clay soil is heavy, smooth, somewhat sticky, and high in nutrients. It tends to hold water well and is relatively slow to drain. Loamy soil is a mix of the other soil types (sand, silt, and clay). It is rich in nutrients and organic matter, drains well, and is generally considered ideal for growing plants.

UNDERSTANDING LIGHT LEVELS

When you're planning a garden, it's important to keep light levels in mind. Some plants need more direct sunlight than others. Others do just fine (or even prefer) shadier sites. Start by observing how many hours of direct, unfiltered sunlight your proposed garden site receives in the summer.

Plants requiring **full sun** will thrive in sunny locations that receive at least 6 hours of full sunlight per day. While such plants may still grow in locations that have less light, they might not do as well or flower as much. Plants characterized as needing **partial sun** or **partial shade** usually do best with 4–6 hours of direct sunlight or dappled light a day. They often thrive when exposed to morning sun and less illumination during the hottest times of the afternoon. Plants requiring **full shade** need fewer than 4 hours of direct sunlight. They often do quite well in locations with dappled shade and prefer direct sun in the morning or the evening. By paying attention to your area's light levels, you'll make your garden as welcoming to pollinators as possible.

Purple coneflowers do best in full sun.

Types of Pollination

Pollination is an important part of reproduction for most plants. It helps them produce a seed that can germinate (sprout) to create new offspring and ultimately continue the species. Pollination happens when individual pollen grains are moved from the anther to the stigma. This act enables fertilization and the later production of fruits and seeds. In most cases, this can only happen between flowers of the same plant species.

Two types of plant pollination can happen. **Self-pollination** is the transfer of pollen within one flower or between flowers on the same plant. **Cross-pollination** is different. It involves the movement of pollen from the flower of one plant to the flower

A bumblebee on an aster flower

of another, either on the same plant or on different plants. Pollen can be transported by gravity (falling off the stigma) or a vector (agent) such as wind, water, or an organism. Most flowering plants need help to move pollen. Plants that use wind tend to produce huge quantities of small, lightweight pollen grains to help guarantee that some will land on another flower. They also tend to produce flowers early in the season before their leaves fully open. This strategy helps ensure that flowers are not blocked by leaves, which would get in the way of the spreading pollen. Wind-pollinated flowers also tend to be small, inconspicuous (not colorful or showy), and lack fragrance and nectar.

Pollen from birch tree catkins being spread by the wind

Animals that transfer pollen are called pollinators. Plants that use pollinators put out smaller amounts of pollen that tends to be sticky or otherwise able to attach to an organism, making it easier to transport. Animal-pollinated plants tend to be colorful, fragrant, and produce nectar to attract and reward pollinators.

HOW POLLINATION WORKS

Corn plants have both male and female flowers. The spiky-looking male flower, called the tassel, has many anthers. It sits at the top of the plant to help wind easily carry and broadly scatter its small pollen grains. A single corn tassel can produce millions of individual pollen grains. Lower down on the plant is the emerging ear. It has numerous silks (long, shiny, thread-like structures) that are the functional stigmas of the female flower. Each individual silk connects to an individual ovule (future corn kernel) on the developing ear. Pollen grains need to land on the silks to successfully pollinate them and fertilize each ovule. The ovule is the organ that develops into a seed after fertilization, which in turn develops into an individual corn kernel. Each fully developed ear of corn can have up to 1,000 kernels or more.

Tassels on a corn plant

Up to 90% of all flowering plants on Earth use animals for pollination. This is critical in maintaining diverse, natural systems; producing the food that we eat; and sustaining much of the life on Earth. While birds, bats, lizards, mammals, and even some marsupials (like opossums) are known pollinators, most pollination is done by insects.

Animal pollinators visit flowers to feed on or collect food. In most cases, this is sugary sweet nectar, but

A brown-belted bumblebee in flight

it may also be pollen. During the visit, pollen from the flower rubs off and sticks to the organism's body. When it moves to another flower of the same species, some of the pollen on the organism's body rubs off onto the flower's stigma, triggering fertilization and completing the process of pollination.

Plants Today

There are anywhere between 300,000 and 450,000 species of plants on Earth. Of these, most can be found in the tropics (warm areas near the equator) and are flowering plants. Together, there are more than 600 plant families, with the Orchid family (Orchidaceae) and Sunflower family (Asteraceae) having the

Pink lady slipper, an orchid

most species worldwide. Within this broader mix are bizarre carnivorous (meat-eating) plants, like the Venus flytrap, which catch and digest small organisms such as insects; an aquatic species called duckweed *(Wolffia globosa)* that is about 1 millimeter in size and considered to be the smallest plant in the world; and the coastal redwood, which can reach heights of over 350 feet. It also includes short-lived species that complete their life cycle in only a month or two, a Great Basin bristlecone pine that was estimated to be nearly 5,000 years old, and even a Mediterranean seagrass colony that some researchers believe may be almost 200,000 years old.

A bristlecone pine

Major Groups of Pollinators

Bees, especially the introduced (not native) western honeybee (*Apis mellifera*), also known as the European honeybee, are well-known pollinators and get a lot of attention. However, many other insects also visit flowers and serve as pollinators. The most common groups include bees, butterflies, moths, flies, wasps, and beetles.

BEES

Bees belong to the insect order Hymenoptera, which also includes wasps and ants. There are almost 4,000 bee species in North America and about five times that many in the world. They come in a wide range of colors and sizes. Some live alone, while others create large colonies. Except for the western honeybee and a few other introduced species, most are **native,** meaning they naturally occur in our area. Beyond visiting flowers for nectar, they also actively collect pollen. These are the primary food resources for adult bees and their developing young. As a result, bees are hardworking and will

A common eastern bumblebee

collect pollen from lots of flowers, often carrying the pollen a long way.

The hairy bodies of bees help them gather pollen when they visit a flower. In fact, the hairs of many insects build up a positive electric charge as they fly and move, much like what you might notice when you shuffle your feet across a carpet and then touch another person or a metal object. As pollen is negatively charged, it is attracted to the positively charged hairs, so bees can also collect pollen without even trying. When bees comb their bodies, the pollen gets packed onto special hairs called **scopae** *(say it, scope-ay)* that are adapted to carry it. A few bee species, such as honeybees and bumble-bees, have a **corbicula** *(say it, core-bick-you-la)*. Also called a **pollen basket,** it is a hollowed-out

Corbicula (pollen basket) on a honey bee

place on the hind leg, surrounded by hairs, where the bee packs its pollen.

BUTTERFLIES

Butterflies are some of the most well-known insects. They tend to be large, colorful, and easy to spot. Along with moths, butterflies belong to the insect

order Lepidop-
tera. There are
around 19,000
species of but-
terflies on Earth.
About 800 species
can be found in
North America.

Monarch butterfly

All adult butterflies
feed on liquids. Most
species in the US and Canada
consume sugar-rich flower nectar.
This energy-rich food helps power their flight and
activity. This is why they love colorful blooming flowers. While feeding, butterflies often brush up against
the flower's anthers and pick up pollen on their heads,
bodies, or wings. Unlike bees, though, most butterflies
do not collect pollen on purpose, nor do they have
special ways to carry it. Nonetheless, butterflies are
still valuable pollinators.

MOTHS

Compared to butterflies, moths are much more
diverse, meaning that there are many more species.
In fact, worldwide, there are roughly 14 times more
moth species than butterfly species. Despite this,
moths don't get as much attention. One reason for
this is that many moths are nocturnal, meaning that

they are active at night. Others are active during the day or at dusk and dawn. Like butterflies, many moth species actively visit flowers to feed on nectar and pick up pollen without trying. When they then visit another flower of the same species, some pollen may rub off and result in pollination. Science suggests that moths play an important role in plant pollination, helping grow many of the food crops that feed us.

A hummingbird moth *(Hemaris)*

Butterfly and moth caterpillars, also called larvae, have chewing mouthparts and feed on plants. More specifically, they feed on certain plants called host plants. These are specialized plants on which female butterflies and moths lay their eggs and which serve as food for the developing larvae. Each butterfly and moth species can only use a certain host plant or group of host plants. For example, monarch butterfly caterpillars can only feed on various species of milkweed *(Asclepias)*. They cannot eat or survive on any other plants. Most butterfly and moth host plants are **native.** These are plants that occur naturally in a particular area or region.

FLIES

A flower fly

Flies belong to the insect order Diptera. With some 17,000 different species in North America, flies are a large and diverse group. They come in a wide variety of sizes, colors, and patterns, with some even looking like bees or wasps. Many, including even some mosquitoes, visit flowers and feed on sugar-rich nectar and sometimes pollen. Like bees, flies tend to have hairy bodies that naturally collect pollen when they feed at flowers. Although easily overlooked, flies are considered important pollinators of wild and crop plants, including cherries, apples, strawberries, raspberries, and many others. Also, the larvae of several species are important in the environment as predators that help provide natural pest control or as decomposers, helping to break down dead plant and animal material.

WASPS

Together with bees and ants, wasps belong to the insect order Hymenoptera. They are the third-largest group of insects. Despite often being feared, most wasps live alone and do not actually sting. By contrast, social wasps that live in groups, like hornets, yellow jackets, and paper wasps, will actively defend their nests and can be very aggressive. They can deliver a painful sting, so be careful not to disturb their nests. It's also a good idea to never handle wasps. Overall, wasps are highly beneficial (good for the Earth). Many are important pollinators and visit a wide variety of flowers. The adults are

A northern paper wasp

equally good predators or parasites of other insects, including many pest species.

BEETLES

A goldenrod soldier beetle

Beetles are the most diverse insect group. They belong to the insect order Coleoptera. In North America, there are around 28,000 species, which is only 7% of the total around the world. Beetles are the largest group of animals, although not all beetles visit flowers. For those that do, they visit flowers in search of food. Most feed on pollen, but some also munch on various flower parts and, less often, feed on nectar. As beetles tend to be large, somewhat clumsy insects, they need to land on and crawl across the flower to feed. As they move through the flower, pollen often collects on their legs and body.

HUMMINGBIRDS

There are around 350 hummingbird species worldwide, all of which are found throughout North, Central, and South America. Most are tropical, with only around 25 species that have been recorded in the US and Canada. Hummingbirds are colorful and highly energetic little birds. They actively visit flowers, hovering at each blossom and using their long, slender beak and tongue to sip nectar. Hummingbirds have a high metabolism (they burn energy quickly) and need to visit many flowers to take in enough food, moving pollen around in the process and serving as important pollinators of many plants. In other areas of the world, several more bird species, such as honeycreepers and even some parrots, feed on nectar and help pollinate flowers.

A ruby-throated hummingbird on penstemon

A bat pollinating a banana flower in South America

BATS

In many tropical or desert areas, bats are important pollinators. In fact, more than 500 different flowering plant species rely on bats for pollination. These include mango, banana, and several species of cactus. Bat-pollinated flowers are typically large, bloom at night, are white or light-colored, and are often quite fragrant. When they visit a bloom, bats often feed on nectar, pollen, and some flower parts. Compared to insects, bats are large and can travel long distances to feed.

Plant-Pollinator Interactions

Flowering plants have evolved alongside their animal pollinators for millions of years. This is an example of **mutualism**—a type of relationship in which both species benefit. The partnership has resulted in a wide variety of ways to ensure that pollination takes place. These include visual cues (color, pattern, or shape), fragrance or scent, food rewards (pollen, nectar, or even fatty oil in some species), mimicry (copying), or even capturing the organism (called entrapment).

VISUAL CUES

Most flowers are showy and attractive. While we enjoy their beauty, these bold displays and bright, colorful petals or sepals are meant to attract pollinators. Think of them like miniature billboards that help make it easier for hungry bees, hummingbirds, and other pollinators to find food in the landscape. Some flowers also evolved **nectar guides.** These are markings, stripes, or patterns, often of contrasting colors, that help lead pollinators to the location of a food reward in the flower, such as sugar-rich nectar. Unlike humans, many insect pollinators, such as bees, can see ultraviolet (UV) light. In other words, bees can see some things in a way that humans can't.

A comparison of a flower in visible light (left) and near-ultraviolet light

The overall shape and size of a flower is another important feature that often guides certain pollinators to access the food rewards. In many cases, the flower shape also increases the likelihood that a visiting pollinator will come in

Coral honeysuckle

contact with the anthers to make sure the pollen gets transferred. While many pollinators can visit different types of flowers, here are some common examples.

Flowers that are shaped like long tubes, pendulous (hanging or downward facing), and those with nectar spurs (long parts that hold nectar) are often pollinated by hummingbirds or sphinx moths that can hover or flutter below the flower and reach the nectar with their long beaks or tongues. A few good examples include red columbine *(Aquilegia canadensis)*, Turk's-cap lily *(Lilium superbum)*, and coral honeysuckle *(Lonicera sempervirens)*.

Turk's-cap Lily

Beyond being long and tube-shaped, many hummingbird-pollinated flowers are funnel-shaped and brightly colored in shades of red, pink, or orange. These flowers offer lots of nectar for hummingbirds. Bees can't see red, so they're less likely to compete with hummingbirds for high-quality food resources. Bees are fast learners, though, and they tend to go to flowers that offer the best rewards.

Broad or shallow tube-shaped flowers with nectar at the base are often visited by bees. Many have a lower lip that provides a good landing platform. The tube shape forces the bees to crawl down the flower to access the nectar, ensuring contact with the anthers. Many species of beardtongue *(Penstemon)* are good examples. Other common traits of bee-pollinated flowers

A Hunt's bumblebee on beardtongue

include a sweet scent or fragrance, bright colors, and being open during the daytime when bees are actively searching for food.

Evening primrose

Moth-pollinated flowers tend to be highly fragrant, are white or pale in color, and are open at night or at dusk and dawn when many moths are active. Many are also tube-shaped with lots of nectar. Examples of moth-pollinated flowers include moonflower *(Ipomoea alba)*, evening primrose *(Oenothera)*, and flowering tobacco *(Nicotiana)*.

Flies visit a wide variety of flowers but are often drawn to shallow, open flowers that are white, purple, or somewhat tan or beige. Many give off an unpleasant odor and smell like rotting fruit or a dead animal and often lack nectar, producing only pollen. Red trillium *(Trillium erectum)* and common pawpaw *(Asimina triloba)* are some common examples.

Red trillium

Common pawpaw

Beetles are not the most graceful of insects. Many tend to be large, heavy, and somewhat clumsy fliers. As a result, they need a fairly large landing platform. Many beetle-pollinated flowers are bowl-shaped or have flattened bunches of many small flowers with easy access to food and often places to hide. Many tend to be pale or green in color, offer abundant pollen for food, and give off strong fragrances ranging from sweet to fruity. The large white blooms of southern magnolia *(Magnolia grandiflora)* are very attractive to beetles.

Magnolia flower

Butterflies often visit a wide range of nectar-rich flowers. Many do, however, prefer flowers with bright colors like red, pink, orange, and yellow, as well as those with nectar guides. They also benefit from daisy-like flowers or clusters of flowers that provide a reliable platform on which to land and feed. Several examples of butterfly-preferred flowers include phlox *(Phlox)*, thistle *(Cirsium)*, milkweed *(Asclepias)*, purple coneflower *(Echinacea purpurea)*, and wild bergamot *(Monarda fistulosa)*.

Great spangled fritillary on thistle

SPECIALIZED RELATIONSHIPS

Many insects are generalist pollinators. They are not super picky and can visit a lot of different flowering plant species that offer food rewards. Specialist pollinators are closely tied with just one or a few certain flowering plant species. They rely only on each other. This tight relationship is an example of **coevolution,** where changes in one organism cause changes in the other organism—and it ultimately increases the chances that pollination will occur.

Ghost orchid

Examples of Unique Relationships or Evolutionary Adaptations

The ghost orchid *(Dendrophylax lindenii)* is found in only a few places in southern Florida, making it one of the rarest flowering plants in the United States. The showy flowers of this mostly leafless plant are pale green to white. Each has two curved pieces hanging from the lower petal and a very thin, long (often over 5 inches) nectar spur hanging off the back. They rely entirely on long-tongued hawk moths for pollination. These nighttime pollinators are the only organisms that can reach the food at the base of the flower's nectar spur.

Some flowering plant species have flowers with anthers that release pollen only through small pores or slits. As a result, their pollen can't be reached by just any flower-visiting insect. Certain bees, including bumblebees, however, have a special strategy called **buzz pollination.** When visiting a flower, they hold the anthers and strongly vibrate their wing muscles, causing pollen to be released. This benefits the plant by reducing pollen loss and increasing its overall chance of pollination. Examples of buzz-pollinated plants include blueberry, tomato, cranberry, and potato.

A close relationship exists between the yucca plant *(Yucca)* and small whitish yucca moths. In fact, they depend completely on one another for survival. After mating, female yucca moths collect the pollen from several flowers. They pack it into a ball and carry it to a flower on another plant, where it's placed on the stigma, thereby fertilizing the flower. This ensures cross-pollination. Finally, the female moth then lays eggs in the flower. When the caterpillars hatch, they feed on the flower's seeds.

Yucca moth on a yucca

Death camas is the common name for a group of plants in the genus *Zigadenus*. All parts of each plant are very toxic or even deadly if eaten—a solid defense from hungry herbivores (plant eaters). Surprisingly, though, even its pollen and nectar are toxic. This unusual strategy helps the plant because generalist pollinators typically visit

Death camas

a wide range of different flowers, which lowers the chance of successful pollination for any one plant species. As a result, death camas relies only on one bee species *(Andrena astragali)* for pollination. This bee and its larvae are not harmed by the toxic chemicals found in the pollen and nectar.

Some plants even use tricks to attract pollinators. Jack-in-the-pulpit *(Arisaema triphyllum)* is a woodland wildflower pollinated by fungus gnats, which are small flies that lay their eggs on fungi and decaying vegetation. The odd-looking hooded flowers of the plant give off a musty, mushroom-like odor. This tricks the gnats into visiting the flower, where they get trapped inside. If it turns out to be a male flower, the gnats eventually escape through a small opening at the base, but not before they are covered in pollen. As

Jack-in-the-pulpit fruit

the gnats continue their search, they will eventually get tricked yet again by a female flower. Once inside, they will work to escape and brush off pollen in the process, thereby fertil-izing the flower. Unfortunately, the small flies will die in the process, as female flowers offer no way out. It is a highly specialized but deadly relationship.

Jack-in-the-pulpit

Why Native Plants Are Important Now

Native plants occur naturally in a particular area or region. As a result, they are adapted to the local environment—meaning that they are used to the soils, rainfall patterns, and temperatures of that location. Native plants are the foundation of most natural ecosystems. They help lessen the impacts of extreme weather, such as droughts or floods; reduce the spread of invasive species; clean our air and water; reduce risk and damage from wildfires; and provide food, shelter, and habitat for countless other organisms.

Over the past 300 years, humans have greatly changed the landscape. Rapid growth of the US population has produced large cities and farms that have taken over natural habitats and resources,

The historic range of prairies in the US

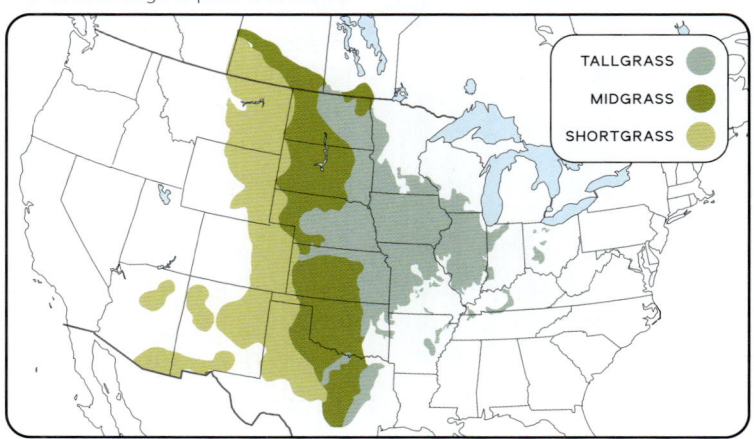

TALLGRASS

MIDGRASS

SHORTGRASS

hundreds of millions of acres in total. For example, about 50% of wetlands, 80% of old-growth forests, and 40% of grasslands have been lost since early European settlement. Today, we lose nearly 150 acres of natural habitat every hour in the United States, meaning animals and plants are running out of places to live.

Sadly, this trend is happening worldwide. **Biodiversity,** the total variety of life on Earth, is being lost at an alarming rate due to growing threats. These include habitat loss, pollution, overuse of pesticides and other chemicals, invasive species, and climate change. Together, this loss of organisms (plants, birds, mammals, insects, and many others), along with their relationships and their many critical roles (such as pollination, natural pest control, and decomposition), puts biodiversity at risk. Many scientists estimate that at least 1 million plants and animals are now at risk of extinction due to human-caused threats—resulting in a **biodiversity crisis.**

If we look a little closer to home, the areas where humans live often look much different than natural environments. Our yards

Periwinkle is a beautiful flower, but it's not native to the US and is often invasive where it is found.

A yard with native plants is far more beneficial for pollinators than a lawn.

and neighborhoods have a lot of **ornamental plants.** These are species of plants that are typically **non-native,** meaning they are living in an area or region where they do not naturally occur, and they're grown mostly for decorative purposes. Some may also be **invasive species.** These are species of animals or plants that are non-native and, once introduced, can aggressively spread, causing harm to native species and the environment. These are sometimes introduced by accident, sometimes on purpose. Many invasive non-native species are sold at nurseries or garden centers, which is the most common way they end up in our neighborhoods. There are more than 4,500 invasive species established in the US, of which about 1,200 are plants.

Lawns are a common feature of many homes. Despite being green, lawns are a mass planting of just one type of grass, typically a non-native species. Unlike diverse natural systems, they provide little or no food or resources for pollinators or other wildlife. They, like many non-native species, also require a lot of water to keep them growing. To keep them healthy and free of weeds, lawns also need chemicals such as pesticides, herbicides, and fertilizers. And lawns need regular mowing. All this effort costs a lot of money, pollutes the environment, and harms or even kills beneficial species such as pollinators and other insects.

Native species, however, are adapted to the soils, weather, and organisms of the area or region where they naturally occur. As a result, they typically have fewer pest or disease problems, require less water and overall care once established, and support other native species. If you like butterflies and moths, for example, the vast majority of the host plants on which their caterpillars feed are native.

Unfortunately, lawns don't provide much benefit for many native plants and animals.

THE PROBLEM WITH CULTIVARS

Cultivars are types of plants that have been bred or grown by humans to have a special feature, such as a flower or leaf color, disease resistance, or size. This is why plants at garden centers often look different (or have different colors) than those found in the wild. Unlike native plants, many cultivars do not benefit pollinators, and some can even be harmful. Some cultivars offer no nectar or pollen resources or produce fruit with less nutrition. Others have showy flowers with additional layers of petals that can make it difficult or impossible for pollinators to reach the available nectar or pollen. The colorful foliage of a few may even be toxic to certain caterpillars or other insects if eaten. While more research is needed to understand how many cultivars are used by, benefit, or even harm pollinators, the best rule is to simply stick with native plants.

A cultivar of cutleaf coneflower

A native cutleaf coneflower
(Rudbeckia laciniata)

Cultivar coneflower

Full Sun
PERENNIAL

Eutrochium maculatum
'Jumpin' Joe'
Joe-Pye Weed

Cultivars are often mistaken for true native plants, but they have been cultivated by people to have particular features (such as brighter colors or bigger flowers). Cultivars have a cultivar name, often in quotes, on their tags.

Full Sun
PERENNIAL

Eutrochium maculatum
Joe-Pye Weed
Native

Native plants do not have cultivar names, and their labels often specify that they are native plants.

HOW TO IDENTIFY CULTIVARS

While it may be challenging to tell a native plant from a cultivar by sight, it's easy if you look at the name. Native plants have a typical scientific name that includes the genus and species and is presented in italics, such as *Eutrochium maculatum*. Cultivars, however, have a non-italicized and capitalized name that is in single quotation marks following the scientific name, such as *Eutrochium maculatum* 'Gateway.' All cultivars have these additional given names.

THE GOOD NEWS: YOU CAN MAKE A DIFFERENCE

Despite the many challenges facing our natural world, every individual and family can help. An easy way to start is to plant a small native plant or pollinator garden. Whether at home, at school, or even as part of a community project, you can work with your parents, grandparents, friends, or teachers to develop a plan and start a garden. Even a small garden or container planting can help provide valuable resources for pollinators and other wildlife. At the back of the book, you will also find a variety of garden plans and fun projects (page 123–137).

You can also help pollinators, native plants, and the environment in the following ways:

- Volunteer at a local park or natural area. You can remove invasive species, help restore habitats, work with your community, and learn a lot!

- Limit unnecessary outdoor lighting at night to help reduce harm to nocturnal insects. Remember that many moths are important pollinators!

- Ask your parents or guardians to reduce their pesticide and herbicide use.

- Help provide nesting sites for native bees. These include bare patches on the ground, brush piles, or plants with hollow stems.

- Get outside and explore. Look closely at the plants and flowers around you. See if you can spot visiting insects. Use this book and other field guides or resources to identify what you see. You will be amazed at what you can find.

How to Use This Book

This book provides a helpful guide to identifying pollinator-friendly native plants. Take it with you on a hike, to the plant nursery, or just to your own yard. This book is organized into three main sections. The introduction of the book will guide you through the basics of botany—what a plant is, how plants evolved, how pollination works, and why native plants are important for pollinators such as bees and butterflies. The second section of the book is a field guide to common native plants with a focus on what pollinators they attract or help. In some cases, there are cultivars of that species that aren't as useful (or outright useless) for native species. We've called out those cultivars with an **AVOID** heading. While this may seem confusing, if you pay close attention to the plant tags and labels, it's not hard to tell them apart. (See page 123 for a checklist.) Finally, if you're ready to start welcoming pollinators to your area, we've included a number of simple garden plans to get you started, including a container-garden plan (perfect for apartment balconies), a garden designed to attract bees, and a butterfly garden, among others.

Get to Know Your Native Plants

Wildflowers (page 58) are the most well-known native plants, and for good reason. They are colorful, and showy, and pollinators love them. That's why the field guide section of the book starts with wildflowers. Some wildflowers are especially important, as they attract many different pollinator species or serve as food sources for caterpillars. We've highlighted these "superhero" plants by discussing them first and including more photos of them.

Trees (page 100) provide a host of resources for pollinators, too, including food for larvae, nectar, and shelter. Several trees, including oaks, maples, and members of the cherry family, are especially important for many insects, including pollinators.

Finally, don't overlook shrubs (page 114) when inviting pollinators to a native garden. They can provide everything from nectar and berries to shelter in the landscape.

MILKWEED

Scientific name: *Asclepias syriaca, Asclepias speciosa, Asclepias tuberosa, Asclepias verticillata, Asclepias viridis, Asclepias incarnata*

Where you'll see it: Throughout the US and southern Canada

What it looks like: An upright perennial plant 1–5 feet tall, its leaves and stems have milky sap. The sweet-smelling umbrella-

Common milkweed
(Asclepias syriaca)

like flowers range in color from white to pinks and purples to orange.

Pollinators: Butterflies, bees, and many other insect pollinators; hummingbirds

Over 70 native milkweeds exist, and most prefer sunny locations, but soil conditions vary from wet to dry, depending on the species. Milkweed serves as a host plant for the monarch *(Danaus plexippus)* and queen *(Danaus gilippus)* butterflies.

A must-have for any pollinator garden, most species flower in spring and/or summer. Some are easy to grow from seed or cuttings.

Avoid: Non-native tropical milkweed *(A. curassavica)*, which is harmful to monarch butterflies

Showy milkweed *(Asclepias speciosa)*

Butterflyweed *(Asclepias tuberosa)*

Whorled milkweed
(Asclepias verticillata)

Green antelopehorn *(Asclepias viridis)*

Swamp milkweed *(Asclepias incarnata)*

AVOID

Tropical milkweed
(Asclepias curassavica)

BEE BALM

Scientific name: *Monarda fistulosa, Monarda didyma, Monarda punctata, Monarda citriodora, Monarda clinopodia, Monarda bradburiana, Monarda media*

Wild bergamot
(Monarda fistulosa)

Where you'll see it: Throughout the US and southern Canada

What it looks like: Bee balm is an upright annual-to-perennial plant 2–5 feet tall with green to gray-green fragrant leaves and branched stems. It has dense, rounded heads of tubular, two-lipped flowers, often with colorful bracts. Flowers range in color from red, pink, and cream to white, depending on species.

Pollinators: Butterflies, bees, wasps, and sphinx moths, as well as hummingbirds

A showy, long-blooming wildflower, these compact plants have highly fragrant leaves that smell like mint. Most species prefer full sun to partial shade and dry-to-average, well-drained soils. Generally easy to grow and readily available from nurseries, this annual species is easy to grow from seed.

Scarlet bee balm *(Monarda didyma)*

Dotted horsemint *(Monarda punctata)*

Lemon bee balm *(Monarda citriodora)*

White bergamot *(Monarda clinopodia)*

Eastern bee balm
(Monarda bradburiana)

Purple bergamot *(Monarda media)*

TICKSEED

Scientific name: *Coreopsis lanceolata, Coreopsis tinctoria, Coreopsis tripteris, Coreopsis verticillata, Coreopsis basalis, Coreopsis grandiflora, Coreopsis major*

Where you'll see it: Throughout most of the US and southern Canada

Lance-leaved coreoposis
(Coreoposis lanceolata)

What it looks like: This is an upright annual-to-perennial plant to 3 feet in height, with a few species much taller. The leaves are green and often highly divided into narrow lobes. The flowers are mostly yellow and daisy-like, with darker centers.

Pollinators: Butterflies, bees, and other pollinators

These plants thrive in open, sunny locations. Most are easy to grow and can be started from seed. Some are ideal for container gardens or small garden spaces. Remove old flowers (this is called deadhead-ing) to promote reblooming. They flower from spring into fall, depending on the species.

Avoid: Cultivars, especially those with many extra pet-als (often called double-petaled)

Plains coreopsis (*Coreoposis tinctoria*)

Tall coreopsis (*Coreopsis tripteris*)

Whorled tickseed
(*Coreopsis verticillata*)

Goldenmane tickseed
(*Coreopsis basalis*)

Largeflower tickseed
(*Coreopsis grandiflora*)

Greater tickseed (*Coreopsis major*)

GOLDENROD

Scientific name: *Solidago canadensis, Solidago speciosa, Solidago velutina, Solidago flexicaulis, Solidago hispida, Solidago nemoralis*

Where you'll see it: Throughout the US and southern Canada

What it looks like: An upright perennial to 7 feet in height with long, green leaves and toothed edges, it has elongated, often branched, and somewhat feathery-looking clusters of small, fuzzy yellow flowers.

Canada goldenrod
(Solidago canadensis)

Pollinators: Bees, butterflies, beetles, other pollinators

Most goldenrod species prefer open, sunny areas and dry-to-average, well-drained soils. They flower from midsummer through late fall. Easy to grow, some species can spread quickly and are somewhat weedy. Goldenrods are an important late-season (end of summer) resource for many pollinators, including migrating monarch butterflies.

Showy goldenrod *(Solidago speciosa)*

Velvety goldenrod *(Solidago velutina)*

Zigzag goldenrod
(Solidago flexicaulis)

Hairy goldenrod *(Solidago hispida)*

Gray goldenrod *(Solidago nemoralis)*

PHLOX

Scientific name: *Phlox drummondii, Phlox paniculata, Phlox diffusa, Phlox divaricata, Phlox multiflora, Phlox pilosa, Phlox subulata*

Where you'll see it: Throughout the US and southern Canada

What it looks like: An upright-to-spreading annual or perennial to 5 feet in height, it has variable, elongated, and narrow-to-needle-like

Spreading phlox
(Phlox diffusa)

green leaves. It produces clusters of fragrant, five-petaled white, purple, pink, blue, or reddish flowers.

Pollinators: Bees, butterflies, sphinx moths, and hummingbirds

There are over 65 native species, and most prefer sunny conditions and average-to-rich, well-drained soils. Some are compact and spread along the ground. Others are tall and can form larger, bushier clumps. Depending on the species, the bloom times vary from spring through summer.

Wild sweet William *(Phlox divaricata)*

Garden phlox *(Phlox paniculata)*

Annual phlox *(Phlox drummondii)*

Flowery phlox *(Phlox multiflora)*

Downy phlox *(Phlox pilosa)*

Moss Phlox *(Phlox subulata)*

SAGE

Scientific name: *Salvia apiana, Salvia greggii, Salvia lyrata, Salvia azurea, Salvia leucophylla, Salvia coccinea*

Where you'll see it: Throughout the US

What it looks like: An upright annual or perennial (occasionally a shrub) to 5 feet in height, it is variable in shape, with fragrant green-to-grayish leaves and square stems. It has clusters of two-lipped tubular white, pink, red, or blue-to-purple flowers.

White sage *(Salvia apiana)*

Pollinators: Bees, butterflies, sphinx moths, and hummingbirds

This is a genus with over 50 native species in the US and a lot of variety between them. Most prefer open, sunny locations, but the preferred soil moisture varies with species. Many kinds of sage are very easy to grow, and it is a must-have for any pollinator garden.

Autumn sage *(Salvia greggii)*

Lyreleaf sage *(Salvia lyrata)*

Azure blue sage *(Salvia azurea)*

Purple sage *(Salvia leucophylla)*

Tropical sage *(Sylvia coccinea)*

San Luis purple sage
(Salvia leucophylla)

BEARDTONGUE

Scientific name: *Penstemon digitalis, Penstemon serrulatus, Penstemon pinifolius, Penstemon cobaea, Penstemon canescens, Penstemon fruticosus, Penstemon hirsutus*

Where you'll see it: Throughout the US and southern Canada

What it looks like: An upright perennial to 5 feet in height, it has somewhat broad-to-narrow

Foxglove beardtongue
(Penstemon digitalis)

leaves of green to gray-green. It produces long clusters of red, purple, pink, or blue-to-white tubular flowers with three lower lobes and two upper lobes.

Pollinators: Bees, butterflies, sphinx moths, and hummingbirds

This is a very diverse genus in North America, with some 280 species found here. Many are available for garden use. Most of them prefer full sun and well-drained soils; it blooms late spring through summer, depending on the species.

Purple cascade penstemon
(*Penstemon serrulatus*)

Pineneedle beardtongue
(*Penstemon pinifolius*)

Prairie penstemon (*Penstemon cobaea*)

Early gray beardtongue
(*Penstemon canescens*)

Bush penstemon
(*Penstemon fruticosus*)

Hairy penstemon (*Penstemon hirsutus*)

SUNFLOWER

Scientific name: *Helianthus annuus, Helianthus divaricatus, Helianthus debilis, Helianthus nuttallii, Helianthus occidentalis, Helianthus tuberosus, Helianthus mollis*

Where you'll see it: Throughout the US and southern Canada

What it looks like: An upright annual or perennial to 10 feet in height, but many species are

Sunflower
(Helianthus annuus)

shorter. It has large and rough-feeling, somewhat triangular-shaped green leaves with toothed edges; its large bright-yellow-to-golden flowers are darker in the center.

Pollinators: Bees, beetles, and butterflies

Sunflowers are a very diverse group of plants, with over 50 native species. Some can be very tall, while others are less than 3 feet in height. Most are easy to grow. They prefer full sun or partial shade and rich, average-to-moist soils. Annual species are easy to grow from seed. Birds eat the seeds in winter.

Woodland sunflower
(Helianthus divaricatus)

Dune sunflower *(Helianthus debilis)*

Nuttall's sunflower
(Helianthus nuttallii)

Fewleaf sunflower
(Helianthus occidentalis)

Jerusalem artichoke
(Helianthus tuberosus)

Ashy sunflower *(Helianthus mollis)*

COLUMBINE

Scientific name: *Aguilegia caerulea, Aquilegia formosa, Aquilegia chrysantha, Aquilegia canadensis, Aquilegia flavescens, Aquilegia longissima*

Where you'll see it: Throughout the US and southern Canada

What it looks like: This is an upright perennial plant 1–3 feet tall with delicate, fern-like green leaves, each with multiple lobed

Colorado columbine
(Aquilegia caerulea)

leaflets. The nodding, bell-shaped flowers have long spurs, and the bloom color varies by species, from red to blue to yellow.

Pollinators: Hummingbirds and bees

Most species prefer partial shade and moist, fertile soils. It can tolerate full sun, however, especially in cooler environments. Often found in open woodlands; along forest edges or streams; and in open, higher-elevation, sunny meadows.

Plants are easy to grow but somewhat short-lived. They are easy to grow from seed. Most species flower in spring and/or summer and are very showy and colorful.

Western columbine
(*Aquilegia formosa*)

Golden columbine
(*Aquilegia chrysantha*)

Red columbine (*Aquilegia canadensis*)

Yellow columbine
(*Aquilegia flavescens*)

Longspur columbine
(*Aquilegia longissima*)

ASTER

Scientific name: *Symphyotrichum laeve, Symphyotrichum ericoides, Symphyotrichum cordifolium; Symphyotrichum novae-angliae, Symphyotrichum foliaceum, Symphyotrichum lanceolatum, Symphyotrichum lateriflorum*

Where you'll see it: Throughout the US and southern Canada

What it looks like: An upright perennial to 5 feet in height, it has green leaves that vary a lot in size and shape. They can be pointed, oval, or even heart-shaped. Asters produce clusters of white, pink, blue, or purple flowers, which are small, daisy-like, and have yellow centers.

Smooth blue aster
(Symphyotrichum laeve)

Pollinators: Bees, butterflies, and other pollinators

A very diverse group with many native species, many are available for garden use. Most prefer full sun to partial shade and average, well-drained soils. It blooms in late summer through fall and provides an important late-season resource for pollinators. Several species serve as butterfly host plants.

White aster *(Symphyotrichum ericoides)*

Blue wood aster
(Symphyotrichum cordifolium)

New England aster
(Symphyotrichum novae-angliae)

Alpine leafybract aster
(Symphyotrichum foliaceum)

White panicle aster
(Symphyotrichum lanceolatum)

Calico aster
(Symphyotrichum lateriflorum)

OHIO SPIDERWORT

Scientific name: *Tradescantia ohiensis*

Where you'll see it: The eastern half of the US and southeastern Canada

What it looks like: Spiderwort forms large, rounded clumps about 2–3 feet tall and just as wide with long grass-like leaves. It has small clusters of blue-to-violet flowers, each with three triangular petals and six bright-yellow anthers. It blooms from spring to midsummer in most areas.

Pollinators: Bees

Spiderwort is naturally found in open, sunny habitats such as prairies, meadows, forest edges, and along roadsides. It prefers full sun and well-drained soil that's dry to lightly moist. It is a good choice for smaller garden spaces or edges.

The colorful flowers open in the morning but last less than a day. These plants provide an abundant source of pollen for bumble bees, honeybees, a variety of smaller solitary bees, and even some flower flies.

EASTERN PURPLE CONEFLOWER

Scientific name: *Echinacea purpurea*

Where you'll see it: The eastern half of the US

What it looks like: This is a sturdy upright plant that is 2–5 feet tall with broad but pointed, rough-feeling green leaves that get smaller up the stem. It has large daisy-like flowers with pink petals and a raised, spiny cone-like center.

Pollinators: Butterflies, bees, and many other insect pollinators, as well as hummingbirds

This showy wildflower is a must for any pollinator garden. It can also be grown in containers. This plant prefers full sun, dry-to-average moisture, and well-drained soil. It is readily available from nurseries. It is easy to grow and can also be started from seed. It blooms from early summer to fall.

The large flowers provide a great landing platform for many insects and access to abundant nectar. It is a particular favorite of many butterflies. After the flowers dry out and fade, they provide seeds for hungry songbirds.

BLANKET FLOWER

Scientific name: *Gaillardia* species

Where you'll see it: Throughout the US and southern Canada

What it looks like: An upright annual-to-perennial plant that is 1–3 feet tall. It has elongated, hairy gray-green leaves, often with lobed edges and large daisy-like flowers with yellow-tipped petals of orange to pinkish-orange and a raised reddish-brown center.

Pollinators: Butterflies, bees, and many other insect pollinators

Very easy to grow, it thrives in full sun and dry-to-average, well-drained soils. It tolerates tough conditions, including heat and drought. It blooms spring to fall. Great for small garden spaces, borders, or even containers, blanket flower can also easily be started from seed.

The brilliantly colored flowers attract many pollinators, especially bees. Several similar-looking species of blanket flower occur in North America.

Avoid: Cultivars (see page 54) of this plant

WILD INDIGO

Scientific name: *Baptisia australis, Baptisia alba*

Where you'll see it: Throughout the eastern US

What it looks like: An upright perennial plant to 4 feet tall, its leaves are green to bluish green and divided into three clover-like leaflets. It features a long cluster of blue, white, cream, or yellow flowers.

Pollinators: Bumblebees, occasionally butterflies and hummingbirds

These bushy, multi-stemmed plants prefer open, sunny conditions and dry-to-average, well-drained soils. They can expand into sizable, shrubby clumps over time. Showy flowers bloom in spring to early summer and are followed by large, inflated seed pods. Wild Indigo is generally easy to grow and requires little care.

BLAZING STAR

Scientific name: *Liatris spicata, Liatris ligulistylis*

Where you'll see it: Eastern two-thirds of the US (east of the Rocky Mountains) and southern Canada

What it looks like: An upright perennial plant to 4 feet in height or slightly more. It has green, elongated, grass-like leaves and a very long, wand-like cluster of small, fuzzy-looking, pink-to-purple flowers.

Pollinators: Butterflies, bees, and other pollinators, including hummingbirds

These plants typically prefer full sun and average-to-moist, well-drained soils. Most species flower from midsummer into fall and flower for a long time. The flower stalks can be up to a foot long or more.

COMMON YARROW

Scientific name: *Achillea millefolium*

Where you'll see it: Throughout the US and southern Canada

What it looks like: An upright perennial to 3 feet in height, this plant has ferny, green leaves that form a low mat. It produces dense, flat-topped clusters of small white flowers on branched leafy stems.

Pollinators: Butterflies, bees, and other pollinators

Wild populations of yarrow in North America represent both native species and introduced plants. It thrives in full sun and tolerates a variety of well-drained garden soils. Low maintenance and easy to grow, it's easy to start from seed. It blooms from summer into early fall, spreads quickly, and does well in containers.

IRIS

Scientific name: *Iris* species

Where you'll see it: Throughout the US and southern Canada

What it looks like: Iris is an upright perennial to 3 feet in height with long, pointed, strap-like green leaves that overlap at the base. Its broad, spreading flowers are mostly violet-blue (but others are white, cream, or even reddish) with short, upward-pointing petals and large, downward-curving sepals that have prominent yellow patches and noticeable dark veining.

Pollinators: Bees and butterflies

Irises are most often found in meadows, forests, or wetlands. They typically grow best in full sun and rich, well-drained soils, but some require extra moisture. Many spread slowly, forming larger clumps or colonies.

JACOB'S LADDER

Scientific name: *Polemonium* species

Where you'll see it: Throughout most of the US and southern Canada

What it looks like: An upright perennial to 3 feet in height, its ferny green leaves are divided into numerous pointed leaflets. It has clusters of bell-shaped flowers that are light blue to purple (occasionally white).

Pollinators: Bees, butterflies, other pollinators

It prefers rich, moist, well-drained soils in partial shade but can be grown in full sun in areas with cooler temperatures. It's a great choice for woodland areas, garden borders, or shadier garden spaces. It blooms in spring to early summer and is a great early-season pollinator plant.

IRONWEED

Scientific name: *Vernonia* species

Where you'll see it: East of the Rocky Mountains and southern Canada

What it looks like: An upright perennial to 7 feet in height, it has long, dark-green leaves with toothed edges. It has wide and somewhat flat clusters of small, fluffy purple flowers.

Pollinators: Bees, butterflies, other pollinators

These plants are tall and fast-growing and prefer sun to partial shade and moist, rich soils. They are great for wooded edges, larger garden spaces, or back garden borders. Flowering midsummer to early fall, they are easy to grow. Some species can spread very quickly.

PASSIONFLOWER

Scientific name: *Passiflora incarnata, Passiflora lutea, Passiflora arizonica*

Where you'll see it: Across the southern half of the US

What it looks like: A perennial vine to 10 feet or more in length, it has palm-like green leaves, typically with three lobes. It uses spring-like tendrils to "climb" (attach to other plants or surfaces). It has showy, highly intricate, green-to-purple flowers.

Pollinators: Bees, butterflies, sphinx moths, and hummingbirds. It is a host plant for several butterflies, including the zebra longwing, the gulf fritillary, the Julia heliconian, and the variegated fritillary.

Passionflowers are primarily a tropical group of plants, and only a few are native to North America. Easy to cultivate, most prefer full sun to partial shade and rich, well-drained soils. Because it is a "climbing" vine, it is best grown on a trellis (a wooden structure) or with some other support. It is a great option for a container garden. Plants bloom almost continuously.

Avoid: Non-native passionflowers, especially red-flowered varieties

SNEEZEWEED

Scientific name: *Helenium autumnale, Helenium bigelovii, Helenium flexuosum*

Where you'll see it: Across the US and southern Canada

What it looks like: An upright perennial to 5 feet in height, it has elongated green leaves. It produces clusters of golden-yellow, daisy-like flowers with domed yellow-to-purplish centers.

Pollinators: Bees, butterflies, and other pollinators

This plant prefers full sun and moist-to-wet, fertile soils. Easy to grow, it forms larger bushy clumps and is great for borders, rain gardens, or in larger groups. Despite its name, it does not cause allergies. It blooms for a long time, from midsummer into fall.

JOE-PYE WEED

Scientific name: *Eutrochium* species

Where you'll see it: Primarily eastern half of the US and southern Canada

What it looks like: An upright perennial to 8 feet in height, it has highly textured green leaves that occur in whorls (the leaves surround the stem) and are spaced out along the smooth, sturdy stem. It produces large, rounded-to-umbrella-like clusters of fragrant, fuzzy pinkish flowers.

Pollinators: Bees, butterflies, and many other pollinators

This plant attracts a lot of pollinators! It forms large multi-stemmed clumps. It prefers full sun to partial shade and moist, rich, well-drained soils. These large plants require space, and they are great for the back of a garden or a border. It blooms from midsummer to early fall.

BLUEBELLS

Scientific name: *Mertensia virginica, Mertensia paniculata*

Where you'll see it: Throughout most of the US and southern Canada

What it looks like: An upright perennial to 3 feet in height, it has smooth, gray-green leaves with clusters of drooping, tubular-to-bell-shaped, light-blue flowers.

Pollinators: Bees, butterflies, sphinx moths, and hummingbirds

This bushy plant does well in sun to shade and moist, rich, well-drained soils. It is perfect for cooler and shadier garden spaces and blooms in spring.

PEARLY EVERLASTING

Scientific name: *Anaphalis margaritacea*

Where you'll see it: Throughout most of the US and southern Canada

What it looks like: An upright perennial to 3 feet in height, it has narrow leaves of gray-green to silvery green, with stem and leaves covered in woolly hairs. It features somewhat flat clusters of small, papery white flowers with yellow centers.

Pollinators: Bees, butterflies, and other pollinators

It thrives in full sun with moist, well-drained soils. It blooms from midsummer into early fall, and its flowers can be dried and used for decoration. Pearly Everlasting serves as a host plant for painted lady *(Vanessa cardui)* and American lady *(Vanessa virginiensis)* butterflies.

ROCKY MOUNTAIN BEEPLANT

Scientific name: *Peritoma serrulata*

Where you'll see it: Throughout the western US (west of the Mississippi River) and southern Canada

What it looks like: An upright annual to 5 feet in height, it has green leaves, each with three pointed leaflets. The pale- to deep-pink flowers each have long, curved anthers. The flowers occur in long, rounded clusters.

Pollinators: Bees, butterflies, sphinx moths, and hummingbirds

Easy and quick to grow from seed, it does best in full sun and average, well-drained soils. Long-blooming, it flowers from early summer through early fall. Plants readily reseed. It can also be grown in containers.

ROSEMALLOW

Scientific name: *Hibiscus moscheutos, Hibiscus coccineus, Hibiscus laevis*

Where you'll see it: Primarily eastern two-thirds of US and southern Canada

What it looks like: An upright perennial to 7 feet in height, it has large green leaves that vary in shape from broad to lobed. It has very large white or pink-to-red trumpet-shaped flowers with spreading petals.

Pollinators: Bees, butterflies, sphinx moths, and hummingbirds

Most species prefer full sun and moist-to-wet soils. It is great for rain gardens, pond edges, or areas of the garden with regular irrigation; the large multi-stemmed plants demand lots of space in the garden. Flowers have a long central stamen that extends out from an often darker center.

GERANIUM

Scientific name: *Geranium maculatum, Geranium oreganum*

Where you'll see it: Throughout the US and southern Canada

What it looks like: An upright perennial to 3 feet in height, it has dark palm-like green leaves with several toothed lobes. It produces small clusters of pink-to-lavender flowers, each sporting 5 rounded petals.

Pollinators: Bees, flies, and occasionally butterflies

Found primarily in meadows or woodlands, it prefers full sun to partial shade and rich, moist, well-drained soils. This plant forms low, mounding clumps and spreads to form larger colonies. It blooms from late spring to midsummer.

CALIFORNIA POPPY

Scientific name: *Eschscholzia californica*

Where you'll see it: Western third of the US and southern Canada

What it looks like: An upright annual (may be perennial in warmer regions) reaching 2 feet in height. It has bluish-green ferny leaves and single, cup-shaped flowers of bright yellow to light orange.

Pollinators: Bees and occasionally butterflies

The state flower of California, it thrives in full sun and dry-to-average, well-drained soils. Easy to grow from seed, it makes a very colorful display in larger groups. It can be grown in containers. A common species in various wildflower seed mixes, it blooms spring to fall.

Caution: Poisonous if eaten

PRAIRIE CONEFLOWER

Scientific name: *Ratibida pinnata, Ratibida columnifera*

Where you'll see it: Eastern two-thirds of the US and southern Canada

What it looks like: An upright perennial with deeply lobed, somewhat fern-like, green-to-gray-green leaves. The flowers are yellow to reddish brown with drooping petals and an elongated, thimble-like center.

Pollinators: Bees, butterflies, and other pollinators

A tall and airy plant, it forms larger clumps and prefers full sun and average-to-dry, well-drained soils. It grows quickly from seed and blooms from early summer to early fall.

MISTFLOWER

Scientific name: *Conoclinium coelestinum, Conoclinium dissectum*

Where you'll see it: South, south-central, and south-western US

What it looks like: An upright perennial to 3 feet in height, it has dark-green leaves that vary from triangular and heavily veined to highly lobed. It produces small clusters of fuzzy light-blue-to-violet flowers.

Pollinators: Bees, butterflies, and other pollinators

This plant thrives in full sun to partial shade and rich, moist soils. It doesn't do well in prolonged dry conditions. This plant is easy to grow and spreads quickly. It can be used as a groundcover and blooms midsummer through fall.

LOBELIA

Scientific name: *Lobelia siphilitica, Lobelia cardinalis*

Where you'll see it: Throughout most of the US and southern Canada

What it looks like: An upright perennial to 6 feet in height, its leaves are green to dark green and vary in shape from narrow to oval, often with toothed edges. The flowers occur in long clusters of white, blue, violet, or red tube-shaped flowers, often with spreading petals.

Pollinators: Bees, butterflies, sphinx moths, and hummingbirds

Grows best in full sun to partially shady sites and rich, moist soils. Most species don't do well with prolonged dry conditions. A great choice for rain or even pond gardens, it can be grown in containers; it blooms midsummer through early fall.

GIANT HYSSOP

Scientific name: *Agastache foeniculum, Agastache scrophulariifolia*

Where you'll see it: Throughout most of the US and southern Canada

What it looks like: This is an upright perennial to 5 feet in height with green, highly veined, fragrant leaves with toothed edges. This plant has a square stem and a short, extended cluster of small two-lipped, tubular light-purple flowers.

Pollinators: Bees, butterflies, wasps, sphinx moths, and hummingbirds

Giant hyssop does well in full sun to light shade and average, rich, well-drained soils. It can handle some drought. Flowers are a pollinator magnet and great for garden beds or containers. It blooms summer through fall.

CHERRY/PLUM

Scientific name: *Prunus emarginata, Prunus ilicifolia, Prunus serotina, Prunus pensylvanica, Prunus angustifolia, Prunus fasciculata, Prunus caroliniana*

Bitter cherry
(Prunus emarginata)

Where you'll see it: Throughout most of the US and southern Canada

What it looks like: Variable in size (from small to large), cherry and plum trees lose their leaves each year. The leaves are dark green, often shiny, with toothed edges. These trees produce clusters of showy white flowers, each with four to five rounded petals, and later small, rounded fruit.

Pollinators: Bees, butterflies, and other pollinators; host plants for several butterflies and moths

Beautiful trees, they also offer wildlife a lot of value. They prefer full sun to partial shade and average, often rich, well-drained soils. Colorful in the fall, birds and many small animals consume the fruit. Flowers bloom in spring. **Caution:** Leaves and seeds are toxic if eaten.

Hollyleaf cherry (*Prunus ilicifolia*)

Black cherry (*Prunus serotina*)

Fire cherry (*Prunus pensylvanica*)

Chicksaw plum (*Prunus angustifolia*)

Desert almond (*Prunus fasciculata*)

Carolina cherrylaurel
(*Prunus caroliniana*)

OAK

Scientific name: *Quercus alba, Quercus rubra, Quercus virginiana, Quercus nigra, Quercus gambelii, Quercus macrocarpa, Quercus chrysolepis*

Where you'll see it: Throughout the US and southern Canada

What it looks like: Oaks are medium-to-large deciduous (or sometimes evergreen) trees with leaves that vary in size and form.

White oak *(Quercus alba)*

Many oak leaves are large and highly lobed. Oak trees also produce male and female flowers on the same tree. Oaks are famous for producing acorns.

Pollinators: The flowers are wind-pollinated, but oaks serve as host plants for nearly 900 different species of moths and butterflies, so they are very important!

A diverse group of trees, there are some 90 species in North America. Oaks offer a lot of value to wildlife. They provide habitat (nesting resources, shelter, etc.) and food for everything from mammals to birds to insects.

Northern red oak *(Quercus rubra)*

Southern live oak *(Quercus virginiana)*

Water oak *(Quercus nigra)*

Gambel oak *(Quercus gambelii)*

Bur oak *(Quercus macrocarpa)*

Canyon live oak *(Quercus chrysolepis)*

MAPLE

Scientific name: *Acer glabrum, Acer macrophyllum, Acer rubrum, Acer saccharinum, Acer saccharum, Acer negundo, Acer pensylvanicum*

Where you'll see it: Throughout the US and southern Canada

What it looks like: Maples are medium-to-large deciduous trees. They have green, palm-like leaves. Their flowers are small and some-

Rocky Mountain maple
(Acer glabrum)

what inconspicuous. Some species are monoecious, while others are dioecious (see page 21). Male flowers look fuzzy and have noticeable stamens. The female flowers typically occur in hanging clusters. Maples flower in spring before leaves have fully emerged. Maples also have winged "helicopter" seeds.

Pollinators: Bees

Wonderful trees in yards and landscapes, maples also provide important early-season resources for bees. Most species prefer full sun and moist, well-drained soils. In the fall, maples produce a lot of color, often changing to shades of yellow, orange, or red.

Bigleaf maple *(Acer macrophyllum)*

Red maple *(Acer rubrum)*

Silver maple *(Acer saccharinum)*

Sugar maple *(Acer saccharum)*

Boxelder *(Acer negundo)*

Striped maple *(Acer pensylvanicum)*

AMERICAN BASSWOOD

Scientific name: *Tilia americana*

Where you'll see it: The eastern half of the US and southern Canada

What it looks like: Basswood is a large deciduous tree with green, somewhat heart-shaped leaves with toothed edges. It produces hanging clusters of small, yellow, highly fragrant flowers.

Pollinators: Bees, flies, and moths. Basswood serves as a host plant for many butterflies and moths.

Basswood is a beautiful landscape tree, and it provides a lot of value for wildlife too. It prefers full sun to partial shade and average-to-moist, rich, well-drained soils. Its seeds are eaten by small mammals.

MAGNOLIA

Scientific name: *Magnolia grandiflora, Magnolia acuminata, Magnolia macrophylla*

Where you'll see it: Throughout the southeastern US north to New York and extreme southern Canada

What it looks like: Magnolia is a medium-to-large evergreen or deciduous tree. It has large, glossy, and somewhat leathery green leaves. Its tulip- to saucer-shaped flowers are large, fragrant, and greenish to creamy white.

Pollinators: Beetles and bees

Most magnolia species are highly ornamental land-scape trees. They thrive in full sun to partial shade and rich, moist soils. They are very easy to grow and maintenance-free. They bloom from late spring into midsummer.

REDBUD

Scientific name: *Cercis canadensis, Cercis occidentalis*

Where you'll see it: Throughout much of the southern two-thirds of the US

What it looks like: A small- to medium-size deciduous tree, redbuds have green, heart-shaped leaves. They have small clusters of pink-to-purple flowers that appear along the branches. Their fruit is a flat, pea-like, hanging brown pod.

Pollinators: Bees, butterflies, and other pollinators

A showy spring species, redbud blooms in early spring before leaves emerge. It prefers full sun to partial shade and average-to-moist, well-drained soils. Birds eat the seeds.

WILLOW

Scientific name: *Salix discolor, Salix exigua, Salix interior, Salix nigra*

Where you'll see it: Throughout the US and southern Canada

What it looks like: Willow is a small-to-large deciduous tree (or sometimes a small shrub). It has elongated, narrow green-to-gray-green leaves and elongated, fuzzy whitish flower clusters (called catkins).

Pollinators: Bees, butterflies, and other pollinators

Willows are a large and diverse genus of plants with around 90 native species. They prefer full sun and moist-to-wet soils and are common in wetlands or along streams, rivers, and ponds. Willows bloom from early spring to early summer, and they are an important early-season source of nectar and pollen. Willows serve as a host plant for many butterflies and moths.

DOGWOOD

Scientific name: *Cornus nuttallii, Cornus alternifolia, Cornus sericea, Cornus canadensis*

Where you'll see it: Throughout the US and southern Canada

What it looks like: An upright deciduous shrub or a small- to medium-size tree. They have somewhat-oval green leaves and fragrant clusters of creamy white-to-greenish flowers, some surrounded by large, modified leaves (called bracts) that resemble petals.

Pollinators: Bees, butterflies, and other pollinators. Dogwood is a host plant for several butterfly and moth species.

Showy and generally easy to grow, dogwood prefers full sun to partial shade and average-to-moist, rich, well-drained soils. It flowers in spring to early summer. The white, red, or black berry-like fruits are eaten by birds and other wildlife.

TULIPTREE

Scientific name: *Liriodendron tulipifera*

Where you'll see it: Eastern half of the US and southern Canada

What it looks like: A large deciduous tree to 100 feet or more, it has large, broad green leaves with pointed lobes. Its large, cup-shaped yellow flowers are marked with orange.

Pollinators: Bees, beetles, flies, and hummingbirds

A common forest tree that is becoming more widely grown in cities. It prefers full sun and moist, rich soils. It blooms from late spring to early summer and is a host plant for some butterflies and moths.

HOLLY

Scientific name: *Ilex opaca, Ilex verticillata, Ilex cassine, Ilex decidua*

Where you'll see it: The eastern half of the US and southern Canada

What it looks like: Ranges in size from an evergreen shrub to a large tree. It has thick, leathery, and often shiny dark-green leaves. Some species have short spines along the leaf edge. Male and female flowers are on separate trees (dioecious, see page 21). It has small greenish-white flowers.

Pollinators: Bees, butterflies, and other pollinators

Holly prefers full sun to partial shade and average-to-moist, well-drained soils. Some species are common landscape plants. It flowers in spring to early summer, and the berry-like black or red fruits are eaten by birds and other wildlife.

A bumblebee on an eastern redbud tree

SERVICEBERRY

Scientific name: *Amelanchier alnifolia, Amelanchier laevis*

Where you'll see it: Throughout the US and southern Canada

What it looks like: An upright deciduous shrub with somewhat-oval green leaves with toothed edges. It has fragrant, star-shaped white flowers with five elongated petals. Its fruit is small, red to black, and berry-like.

Pollinators: Bees, butterflies, and other pollinators, including some hummingbirds. It is a host for caterpillars of several kinds of butterflies and moths.

Grows well in full sun to partial shade and rich, well-drained soils. Flowers appear before the leaves fully emerge. It blooms in spring to early summer. The fruit is eaten by birds and other wildlife, and it provides great nesting sites for birds.

BLUEBERRY

Scientific name: *Vaccinium corymbosum, Vaccinium arboreum, Vaccinium ovatum*

Where you'll see it: Throughout the US and southern Canada

What it looks like: An upright shrub to a small tree with leaves that are small, somewhat oval, and often glossy green. It has clusters of small white, bell-shaped flowers and round, purple-to-black, berry-like fruit.

Pollinators: Bees, butterflies, and other pollinators. It is a host for several butterfly and moth caterpillars.

It attracts a large amount of wildlife. It prefers sun to partial shade and moist, well-drained acidic soils. Relatively easy to grow, blueberry plants bloom in spring to early summer.

CEANOTHUS

Scientific name: *Ceanothus cuneatus, Ceanothus americanus, Ceanothus thyrsiflorus*

Where you'll see it: Throughout the US and southern Canada

What it looks like: An upright deciduous or evergreen shrub, it has oval-shaped, often heavily veined green leaves. It produces rounded clusters of small white, pinkish-to-blue flowers.

Pollinators: Bees, butterflies, and other pollinators

A genus with a lot of variety, these plants are very showy, with some species similar looking to lilac. It prefers full sun and dry, well-drained soils and blooms from spring to early summer.

SUMAC

Scientific name: *Rhus glabra, Rhus typhina*

Where you'll see it: Throughout the US and southern Canada

What it looks like: An upright deciduous shrub, it has oval-shaped, long, fern-like green leaves. It has long upright clusters of small green-to-white flowers and produces fuzzy reddish-to-brown seeds.

Pollinators: Bees, butterflies, and other pollinators. It serves as a host for many butterfly and moth caterpillars.

It prefers full sun to partial shade and average-to-moist, well-drained soils. Some species may spread quickly, forming larger colonies. It produces male and female flowers on different plants (dioecious) and blooms spring to midsummer.

CURRANT or GOOSEBERRY

Scientific name: *Ribes americanum, Ribes aureum, Ribes cereum, Ribes oxyacanthoides, Ribes sanguineum*

Where you'll see it: Throughout most of the US and southern Canada

What it looks like: A small, upright, deciduous shrub, it has 3-5 lobed green leaves with coarsely toothed edges. It produces short clusters of long tubular cream, yellow, or reddish flowers, each with five spreading lobes.

Pollinators: Bees, butterflies, sphinx moths, and hummingbirds

It prefers full sun to partial shade and tolerates a variety of well-drained soils. Easy to grow and care for, it blooms in spring. Numerous songbirds and small mammals consume the fruit, and it provides cover for songbirds and other wildlife.

VIBURNUM

Scientific name: *Viburnum opulus, Viburnum acerifolium, Viburnum dentatum, Viburnum obovatum*

Where you'll see it: Throughout most of the US and southern Canada

What it looks like: A small-to-large deciduous shrub, its green leaves vary in shape. It produces a flat cluster of small white flowers and has red or black berry-like fruit.

Pollinators: Bees, butterflies, and other pollinators. Many songbirds and small mammals consume the fruit. It also provides cover for songbirds and other wildlife

Beautiful landscape plants, virburnums offer a huge amount of wildlife value. Easy to grow, it prefers full sun to light shade and average-to-moist, well-drained soils. It blooms in spring; many species also display attractive fall color.

NINEBARK

Scientific name: *Physocarpus capitatus, Physocarpus opulifolius*

Where you'll see it: Throughout most of the US and southern Canada

What it looks like: A large deciduous shrub with lobed green leaves with coarsely toothed edges. It has small domed clusters of fuzzy white, five-petaled flowers.

Pollinators: Bees, butterflies, and other pollinators

Many songbirds and small mammals eat the seeds, and ninebark provides great cover for songbirds and other wildlife.

A butterfly on a Ninebark shrub

Purple loosestrife

Japanese honeysuckle

English ivy

Japanese barberry

Kudzu

Chinese wisteria

Lantana

Norway maple

Autumn olive

Cogongrass

AVOID INVASIVE SPECIES!

Projects, Activities, and More

KNOW YOUR INVASIVES

Many non-native plants are used in urban areas, such as cities, neighborhoods, and yards. Some can be very attractive to pollinators, providing nectar and pollen. Some even serve as host plants for caterpillars.

Other non-native plants are invasive. Invasive plants are non-native species that cause harm to the environment or the economy. In many cases, invasives outcompete native plants. This can dramatically change habitats, reduce biodiversity, and even cause extinctions. To help prevent the spread of invasive plants, avoid purchasing them from garden centers. You and your parents can also join volunteer efforts to remove them from natural spaces like parks.

BEFORE YOU BUY THAT PLANT

Here's a quick checklist to make sure you're buying native plants. First, pay attention to where you're shopping. Most big-box home-improvement stores don't carry many native plants. They usually carry ornamental plants, often non-native species from Europe or Asia. If they do carry species found locally, they are often cultivars, not true native species. These plants, while pretty, don't offer many (if any) benefits to native pollinators and species. If these larger stores do offer native plants, they are often found in a special (and usually small) section.

If you want native plants, you often need to find nurseries that specialize in native plants.

When shopping, review plant labels carefully and avoid cultivars. Cultivars usually have the common name of the plant, followed by the cultivar name in single quotation marks. Here's an example of a cultivar label:

Full Sun
PERENNIAL

Eutrochium maculatum
'Jumpin' Joe'
Joe-Pye Weed

After you've found a native plant, make sure that it's a good fit for the area and soil in which you'll plant it.

Finally, make sure the plant wasn't pretreated with insecticides. Some plants, especially from big home-improvement stores, are treated with an insecticide (usually neonicotinoids), which can hurt bees and other pollinators. Native plant nurseries often do not treat their plants with such chemicals.

BASICS OF PLANNING A GARDEN

Starting Plants from Seeds

Gathering seeds from your native flowering plants is a fun and easy way to grow new plants and expand your garden. By collecting locally, you also ensure that the resulting plants are locally adapted to your soils

A variety of wildflower seeds

and climate. Be sure to only collect seeds from your garden or yard and not from wild spaces, as this can harm local plant populations and typically requires a permit. Before collecting from a neighbor's property, get permission first.

Plants begin to set seed after they are done flowering but may take many weeks to fully ripen. Typically, they are ready to harvest when they are brown and are ready to easily fall out of their pod or flower head.

What You'll Need

- Garden clippers to cut old flower heads or seed pods
- Several paper lunch bags
- Permanent marker
- Paper clips
- Zip-top plastic bags

When you locate a plant with harvestable seeds, use the clippers to cut the old flower head or seed pod and place the seeds in the paper bag. Avoid harvesting all the seeds from your plants. Leaving some behind will provide food for birds and other wildlife

 and offer an opportunity for your plants to naturally self-seed. Also, avoid harvesting seeds in wet conditions. Next, label the back with the plant name and date.

Store your seed-filled paper bags in a cool, dry, and dark location. This works well for short periods of time. If you plan to store seeds for longer, such as over the winter, transfer them to zip-top plastic

A wildflower with seeds visible

bags and place them in the refrigerator. Be sure that the seeds are dry; otherwise, mold may form.

When to plant your seeds depends on where you live and what you wish to do. You can simply sow them back in your garden after collection, expanding an area with a particular plant or picking an entirely new location. This is the simplest method and is essentially what nature does. Or you can plant the seeds in a pot or seed-starting tray. This is more work but very rewarding, and it gives you more control over the process. It may be best to do a little research to get recommendations on when and how to plant the seeds of a particular species, as this can vary. Some species may need an extended period of cold temperatures to germinate (stratification), while others benefit from breaking down the seeds' protective coat (scarification). Some species may also benefit from being sown on the surface of the soil versus being covered with soil. Fortunately, there are lots of good resources available (page 142).

Purchasing Seeds

You can also ask your parents or guardians to buy seeds to plant. Annual species are often the easiest way to start, as they don't require pre-treatment (see above) like perennials do and they grow quickly.

Growing Plants from Cuttings

Growing plants from cuttings is another fairly straightforward way to create more plants.

What You'll Need

- A pair of garden clippers or scissors
- A small pot with a saucer
- Potting soil
- A pencil
- Rooting hormone (can be purchased online or from any garden center)
- A plastic bag and a string

First, locate a plant you want to propagate. Pick a plant with new green and non-woody growth. Most perennials (not shrubs or trees) are a good choice. Next, choose a healthy stem and look for a node. This is a small, enlarged area or bump right below a leaf. Using a clean pair of garden clippers or scissors, cut right below the node so that your cutting is about 6 inches long. Next, cut all but two or three leaves, including the leaves nearest the node. Now, your cutting is ready to plant. Fill your pot up with soil and moisten it slightly. Next, tamp down the soil with your hand. Then, using a pencil, make a hole in the soil. Move the pencil around a bit to ensure that the hole is larger than the diameter of your cutting. Moisten the bottom portion of your cutting with water and then dip it in the rooting hormone powder. Gently tap off any excess powder, and then place your cutting into the soil hole. Using your hands, tamp the soil down around your cutting so that it firmly stands up. You can often put several cuttings into the same pot if needed. Next, lightly water the cutting and place the pot in a plastic bag. Pull the open end of the plastic bag together and loosely tie it with string. Do not completely seal the plastic bag, but leave a little opening to allow air to circulate. This creates a little greenhouse to help your cutting germinate. Lastly, place your pot in a warm spot with indirect light. Do not place it in full sun. Check your cutting regularly and add a bit more water if the soil becomes dry. Successful cuttings will remain green and not look wilted. After a few weeks, your cutting should begin to develop roots. To help check, lightly pull on the cutting. If you feel some resistance, your cutting likely has developed roots and is now ready to transplant.

REARING A CATERPILLAR

Woolly bear

Butterfly and moth caterpillars, also called larvae, are often found in yards and nearby wild spaces. They are herbivores (feed on plants). Taking care of them takes time and commitment, but it can be quite fun and rewarding. Doing so gives you an opportunity to enjoy them, see them up close, watch them grow, pupate (or spin a chrysalis or a cocoon), and eventually emerge as an adult butterfly or moth.

What You'll Need

- A regular supply of fresh, high-quality host-plant material (see below)

- A pair of garden clippers to cut plant stems or branches

- A safe, roomy, and secure environment (plastic cups or containers are an option, but a mesh pop-up cage is best)

- A caterpillar (see safety note below)

- A zip-top plastic bag with a little water in it

Note: Don't handle caterpillars, as it's easy to accidentally harm them. Also, some moth larvae can have irritating or stinging spines or hairs. It's best to simply move the caterpillar by cutting the old branch or stem on which the caterpillar is resting, placing it onto fresh plant material, and letting the caterpillar crawl out on its own.

In general, only raise caterpillars that you find on plants. This will ensure that you know what to feed them—more of the plant you found them on. Caterpillars have very specialized dietary needs and are unable to feed on just any plant. They rely on specific host plants to complete development. Once you identify the correct food, you can use either potted plants or freshly cut plant material to feed your caterpillar. If you are using cut plant material, place the stems into a plastic container of water (old plastic soda or water bottles work well). Avoid tightly sealed containers, as they can cause humidity and moisture to build up, which can lead to mold and disease.

Make sure that the opening of the container holding the water does not have a large gap around the plant stems. Otherwise, caterpillars may crawl down into the water and drown. You can use aluminum foil to help create a tighter seal. Then place the container of vegetation in your cage or aquarium and put the caterpillar on the plant material.

Then seal the cage or cover the aquarium to ensure the caterpillar does

Monarch butterfly caterpillar

not escape. Replace the leaves, add new water to the container, and clean out the cage or aquarium on a daily basis. Depending on the species and the size of the caterpillar you find, some caterpillars may take weeks to fully develop. Remember that the caterpillar will die without healthy food and clean conditions.

Monarch butterfly chrysalis

Once the caterpillar is fully grown, it will often crawl off the plant material and find a secure place to pupate or spin a chrysalis (or if it's a moth, a cocoon). Once it has done so, place a few extra branches or sticks in the cage so that the emerging moth or butterfly has something to crawl up on and expand its wings. Cocoons or pupae produced in late summer or fall may overwinter, with the adults not emerging until the following spring. If it is late in the season and the moth or butterfly has not emerged after about 3 to 4 weeks, place the cage in a cool, dark location such as a garage or basement. This should help keep the insect from accidentally emerging until the following spring. Check on them periodically over the winter. Then, when the weather warms, bring them back into the house or onto a porch in a bright location out of direct sunlight. This should encourage the adult moth or butterfly to emerge, and you can observe it for a short time before releasing it into your yard or a nearby wild area.

MILK JUG GREENHOUSE

You can quickly and inexpensively make a neat little greenhouse at home. You can use this to start seeds for your pollinator garden or container garden. This is a great way to learn how to germinate seeds and watch your plants grow. It can be done with most any annuals or perennials, and it allows you to get an early jump on spring.

What You'll Need

- A 1-gallon plastic milk container
- A pair of scissors or a box cutter
- Duct tape
- Potting soil
- Seeds
- A permanent marker

Milk jug greenhouse collection

Using the scissors or box cutter (with an adult's help), make several small holes in the bottom of the milk container. This will allow water to drain out and not collect in the bottom. Next, remove the plastic lid. Again, with an adult's help, using the scissors or box cutter, cut about three-quarters of the way around the center of the container, leaving about 2 inches attached near the handle as a hinge. This will allow the top to flip up and down. As these tools are sharp, it's best to have a parent or other adult help you.

Next, add about 3 to 4 inches of moist potting soil to the bottom. Then, sprinkle the seeds of your chosen plant on the soil. Be sure to follow any directions on the seed packet, as some plant species need their seeds covered with soil, while others are fine with just being sprinkled on top. Once your seeds are added, close the lid and seal it around the middle with duct tape.

A milk jug greenhouse in action

Do not place the plastic top back on. Then label the container with the species of plant and the date.

While this can be done at a variety of times, it is best to start plants early in the spring before the weather gets too warm. You can place your milk jug greenhouse outside in a sunny location and wait for your young seedlings to germinate. The sun will heat up the soil and create condensation, much like a real greenhouse. Check the seedlings periodically, and water them as needed to prevent the soil from completely drying out.

Once the seedlings are up and have a few leaves—and if the weather is warm enough with no chance of a freeze or frost—you can transplant them into your garden or a pot and continue to watch them grow.

CONTAINER GARDEN FOR POLLINATORS

If you don't have access to a yard but want to attract bumblebees and other native species, you can plant a container garden. When situated on a balcony or a patio, they can bring nature to you!

Swamp milkweed

Autumn sage

Black-eyed Susan

Other good container species include: **Blanket Flower** (*Gaillardia*), **Tickseed** (*Coreopsis*), **Purple Coneflower** (*Echinacea purpurea*), **Common Yarrow** (*Achillea mille-folium*), **Phlox** (*Phlox*), **Passionflower** (*Passiflora*), and **Mistflower** (*Conoclinium coelestinum*).

A BASIC BEE GARDEN

If you want to attract bees, providing a good deal of nectar-rich plants is a step in the right direction.

Beardtongue

Yarrow

Black-eyed Susan

Beebalm

Aster

Sunflower

Sneezeweed

Blazing Star

Other good choices include: **Phlox** *(Phlox)*, **Ironweed** *(Vernonia)*, **Wild Indigo** *(Baptisia)*, **Joe-Pye Weed** *(Eutrochium)*, **Sage** *(Salvia)*, **Rosemallow** *(Hibiscus)*, **Lobelia** *(Lobelia)*, **Giant Hyssop** *(Agastache)*, and **Blueberry** *(Vaccinium)*.

A BASIC BUTTERFLY GARDEN

To attract butterflies, it's important to offer plants rich in nectar, but it's equally important to offer flowers that bloom over the course of summer and into fall. That way, your garden will be welcoming no matter when a butterfly visits.

Other good choices include: **Ironweed** *(Vernonia)*, **Sage** *(Salvia)*, **Rosemallow** *(Hibiscus)*, **Lobelia** *(Lobelia)*, **Giant Hyssop** *(Agastache)*, **Passion-flower** *(Passiflora)*, **Prairie Coneflower** *(Ratibida)*, and **Ceanothus** *(Ceanothus)*.

Phlox

Dotted horsemint

Goldenrod

Blue mistflower

Liatris

Joe-Pye weed

Black-eyed Susan

Narrowleaf milkweed

Purple coneflower

Glossary

Angiosperms (an-gio-sperms)   A major group of seed-bearing plants. Also called flowering plants, their seeds are enclosed by a fruit.

Anther   the pollen-producing structure of the stamen

Binomial nomenclature   The formal system of naming organisms. It consists of two parts, the genus and species. For example, a house cat is Felis catus.

Biodiversity   the total variety of life on Earth

Biodiversity crisis   the alarming loss of biodiversity due to an increasing number of human-caused threats

Botanists   people who study plants

Botany   the scientific study of plants

Bracts   modified leaf-like structures below the calyx that can serve to protect flowers, fruits, or seeds from herbivores or cold temperatures

Buzz pollination   a behavior used by some bees to grasp the anthers and strongly vibrate their wing muscles, causing pollen to be released

Calyx   the combined group of sepals of a flower

Chlorophyll   a green pigment found within chloroplasts that gives plants their green color and absorbs energy from the sun

Chloroplasts   specialized structures in plant cells that carry out the process of photosynthesis

Coevolution   a tight relationship, where changes in one organism cause changes in the other organism

Corolla   the combined group of petals of a flower

Cross-pollination   the movement of pollen from the flower of one plant to the flower of another.

Dioecious (die- EE-shuhss)   plants that are either all male or all female. Both a male and female plant are needed to ensure reproduction and produce fruit.

Filament (fil-uh-ment)   an elongated stalk-like structure of the stamen that supports the anther

Fruit   the fertilized, mature ovary that contains one or more seeds

Genus (Gee-nuhss)   the first part of a scientific name. It indicates which organisms are closely related.

Gymnosperm (jim-no-sperm)   A major group of seed-bearing plants. They do not produce flowers and produce naked seeds that are not enclosed in a fruit.

Inflorescence (En-floor-es-ence)   the arrangement of flowers on a stem

Invasive species   species of animals or plants that are nonnative and accidentally or intentionally introduced to a given location. Once introduced, they can aggressively spread, causing significant harm to native species and the environment.

Monoecious (mon-EE-shuhss)   plants that produce separate male and female flowers on the same plant

Mutualism (myoo-choo-uh-liz-uhm)   a type of relationship in which both species benefit

Native   animals or plants that occur naturally in a particular area or region

Native plants   plants that occur naturally in a particular area or region

Nectar guides   markings, stripes, or patterns, often of contrasting colors, that help lead pollinators to the location of a food reward in the flower, such as nectar.

Nonnative   species of animals or plants that are living in an area or region where they do not naturally occur

Ornamental plants   species of plants that are typically nonnative and grown mostly for decorative purposes

Ovary   the enlarged base of the pistil. Once fertilized, the ovary develops into a fruit containing one or more seeds.

Petals   modified leaves that surround the reproductive part of a flower. They are often brightly colored to attract pollinators and together make up the corolla.

Pedicel   an individual stalk that supports a flower

Photosynthesis (pho-to-syn-the-sis) the process by which green plants transform light energy from the sun, water taken up by roots, and carbon dioxide from the atmosphere to fuel a chemical reaction that results in the production of sugar (glucose) and oxygen

Pistil the female part of a flower, it can be found in the center of the flower.

Plantae (Plan-tae) one of the kingdoms of life that include all plants

Pollen fine, grainy substance produced by the male part of a flower. It contains the male reproductive cells and is typically transported by wind, water, or an organism.

Pollen basket a cavity-like structure on the hind leg of a bee that is surrounded by hairs and into which pollen is packed for transport

Receptacle a somewhat enlarged section or foundation of the pedicel to which all the various flower parts are attached

Scopae (Skow-pay) specialized masses of dense, branched hairs that are designed to carry pollen

Seeds a small, mature structure produced by the reproductive parts of a plant that can produce a new plant

Self-pollination the transfer of pollen within one flower or between flowers on the same plant

Sepals (Se-puhl) the outermost parts of a flower that help protect the flower bud and eventually the developing fruit

Species the second part of a scientific name. It identifies a group of individuals with similar characteristics that can produce fertile offspring.

Stamen (Stay-muhn) the male reproductive part of a flower

Stigma a somewhat flattened or bulb-like structure at the top of the style

Style an elongated structure supported by the ovary that turn is topped by a somewhat flattened or bulb-like structure called a stigma

Taxonomy the science of naming and classifying organisms

Recommended Reading

ONLINE RESOURCES

Seek by iNaturalist (inaturalist.org/pages/seek_app). This application uses the camera on your smartphone or tablet, along with image recognition, to help you identify insects, plants, and other animals.

BugGuide (bugguide.net/node/view/15740). An online resource providing identification, images, and information on insects, spiders, and their relatives for the United States and Canada

Xerces Society for Invertebrate Conservation's Pollinator Conservation Resource Center (xerces.org/pollinator-resource-center). An online resource for region-specific collections of publications, native plant and seed suppliers, and other resources to aid in planning, establishing, restoring, and maintaining pollinator habitat

Pollinator Partnership's Pollinator Conservation Resources (pollinator.org/pollinator-resources). An online resource to help you get started with your pollinator conservation efforts

Bumblebee Watch (bumblebeewatch.org). A program designed to track and conserve North American bumblebees, it relies on information collected by community scientists, and you can help!

FIELD GUIDES AND OTHER BOOKS

Daniels, Jaret. *Insects & Bugs for Kids: An Introduction to Entomology*. Adventure Publications, 2021.

Daniels, Jaret. *Backyard Bugs: An Identification Guide to Common Insects, Spiders, and More*. Adventure Publications, 2017.

Daniels, Jaret. *Native Plant Gardening for Birds, Bees & Butterflies*. A series by Adventure Publications with books for the Upper Midwest, Lower Midwest, Northeast, Pacific Northwest, South, and Southeast.

Miller, George. *Native Plant Gardening for Birds, Bees & Butterflies*. A series by Adventure Publications with books for the Rocky Mountains and the Southwest.

Mizejewski, David. *National Wildlife Federation®: Attracting Birds, Butterflies, and Other Backyard Wildlife, Expanded Second Edition*. Design Originals, 2019.

The Xerces Society. *Attracting Native Pollinators: The Xerces Society Guide, Protecting North America's Bees and Butterflies*. Storey Publishing, LLC. 2011.

Diane Woodcheke: 48; **Doikanoy:** 10 (Water oak), 13 (Sunflower); **Donna Bollenbach:** 8, 94, 101 (Black cherry), 105 (Striped maple); **Edita Medeina:** 54 (both), 124; **Esa Hiltula:** 114; **Evannovostro:** 137 (Purple coneflower); **FatimeBarut:** 122 (English ivy); **Flower_Garden:** 122 (Purple loosestrife); **Francisco Blanco:** 44; **Francisco Herrera:** 69 (Tropical sage); **Gabriela Beres:** 65 (Gray goldenrod), 76; **Gerry Bishop:** 14 (top right), 40 (bottom), 59 (Swamp milkweed), 85, 134 (middle); **Gonzalo de Miceu:** 79; **GraphicsRF.com:** 10; **guentermanaus:** 9 (Genus), 63 (Plains coreopsis), 69 (Azure blue sage); **Gurcharan Singh:** 10 (Hollyleaf cherry); w**Hanahstocks:** 6 (Mosses); **Hillside Studios:** 73 (Dune sunflower); **homi:** 122 (Kudzu); **Hun Young Lee:** 74; **Igor Klyakhin:** 26; **iPlantsman:** 111; **Irina Borsuchenko:** 105 (Silver maple); **Iris - Beautiful Nature:** 106; **Isabel Eve:** 103 (Southern live oak); **Iva Vagnerova:** 15 (flowering plants); **iwciagr:** 122 (Norway maple); **Jacob Tian:** 129 (top); **Jananz:** 53 (bottom left); **Jane Rix:** 42 (bottom); **Jared Quentin:** 101 (Desert almond), 103 (Canyon live oak); **Jennifer Bosvert:** 75 (Yellow columbine); **Jerrold James Griffith:** 61 (White bergamot); **Jim Beers:** 61 (Eastern bee balm); **Jim Schwabel:** 73 (Fewleaf sunflower); **John_P_Anderson:** 87; **Kabar:** 9 (Order), 58, 63 (Largeflower tickseed & Whorled tickseed), 70, 71 (Purple cascade penstemon), 135 (Sneezeweed), 137 (Phylox); **Karel Bock:** 82, 135 (Blazing Star); **Kateryna Mashkevych:** 9 (Class); **Kathryn Atwater:** 59 (Butterflyweed); **Kenneth Keifer:** 90; **Kerrie W:** 131; **Kirsanov Valeriy Vladimirovich:** 28 (bottom); **Kirti Bhole:** 9 (Phylum); **KLLarson:** 116; **KylieP:** 132; **Larry Eiden:** 6 (Shrubs); **LarryDallaire:** 34; **Laurin Rinder:** 69 (San Luis purple sage); **LifeCollectionPhotography:** 122 (Chinese wisteria); **LifeisticAC:** 84; **Linda McKusick:** 28 (top), 36 (bottom); **Lost_in_the_Midwest:** 133; **M. Schuppich:** 95, 109; **Mao-Tung Hsu:** 14 (ferns, middle); **Maple Ferryman:** 134 (right); **mar_chm1982:** 67 (Garden phlox); **marekuliasz:** 125; **Marianne Pfeil:** 63 (Goldenmane tickseed); **Mariola Anna S:** 41 (bottom), 81; **Martina Simonazzi:** 6 (Trees); **Martina Unbehauen:** 14 (horsetail, middle); **mcajan:** 101 (Fire cherry); **Meagan Marchant:** 59 (Tropical milkweed); **Media Marketing:** 24; **Melody Mellinger:** 47; **meunierd:** 61 (Lemon bee balm); **Michael Koenen:** 135 (Black-eyed Susan); **Michael Siluk:** 35; **Michael Stubben:** 77 (Alpine leafybract aster); **MIROFOSS:** 46; **Nahhana:** 107; **Nancy J. Ondra:** 9 (Species), 61 (Dotted horsemint), 63 (Tall coreopsis), 98, 137 (Dotted horsemint); **Nathan mamico:** 73 (Woodland sunflower); **Nazaruk Nazar:** 117; **New Africa:** 52; **Nikki Yancey:** 103 (Gambel oak); **Nikolay Kurzenko:** 108; **Nina B:** 59 (Showy milkweed), 71 (Bush penstemon); **nnattalli:** 9 (Family); **nomis_h:** 77 (Blue wood aster); **Oleg Kovtun Hydrobio:** 14 (Freshwater algae); **Olga S photography:** 102; **olko1975:** 22, 115; **Orest lyzhechka:** 135 (Yarrow); **Pascal Guay:** 25; **Paul Reeves Photography:** 31, 36 (top), 99; **Phill Doherty:** 42 (top); **PhotOleh:** 103 (Northern red oak); **Przemyslaw Muszynski:** 6 (Ferns); **Pyty:** 15 (conifer, pinecones); **R. Maximiliane:** 72; **Randy Bjorklund:** 75 (Western columbine), 110; **Ray Hugo Tang:** 6 (Green algae); **Ray vizgirdas:** 100; **Reikara:** 83; **Robert Biedermann:** 7 (Early gray beardtongue); **RukiMedia:** 73 (Ashy sunflower); **Sandiwild:** 6 (Liverworts); **Sergey V Kalyakin:** 67 (Moss phlox); **Sinhyu Photographer:** 13; **SoFlo Shots:** 67 (Flowery phlox); **Somsit:** 122 (Japanese honeysuckle); **somyot pattana:** 73 (Jerusalem artichoke); **Stephen Bonk:** 89, 137 (Joe Pye weed); **Steve Samples:** 43; **Sticky Notes Queen:** 91; **Sundry Photography:** 66, 137 (Narrowleaf milkweed); **Sunshower Shots:** 105 (Red maple); **Super:** 14 (top left); **Susan Hodgson:** 41 (top); **Svetlana Akhmedova:** 23; **tamu1500:** 27, 68; **Tarumaphoto:** 122 (Cogongrass); **Teri Virbickis:** 92; **The Jungle Explorer:** 59 (Green antelopehorn); **Tibesty:** 135 (Aster); **Timothy H Brown Jr:** 77 (Calico aster); **Todd Boland:** 65 (Hairy goldenrod), 75 (Longspur columbine); **Tom Meaker:** 101 (Chickasaw plum), 119; **Tonio_75:** 17; **traction:** 64; **Traveller70:** 6 (Grasses), 88; **tviolet:** 60; **Tyler Wenzel:** 38; **Unkas Photo:** 122 (Japanese barberry); **v.apl:** 21; **vagabond54:** 105 (Bigleaf maple); **Vahan Abrahamyan:** 61 (Scarlet bee balm), 135 (Beebalm); **Veronica Skelton:** 45 (bottom); **watcher fox:** 62; **Will Pedro:** 29; **William Cushman:** 71 (Prairie penstemon); **Wirestock Creators:** 63 (Greater tickseed); **Zerbor:** 16; and **zzz555zzz:** 97.

About the Author

Jaret C. Daniels is a curator at the Florida Museum of Natural History's McGuire Center for Lepidoptera and Biodiversity, and he is a professor in the Department of Entomology and Nematology at the University of Florida. Jaret holds a Ph.D. in entomology and specializes in the ecology and conservation of at-risk butter-flies and other native insect pollinators. He is a professional nature photographer and author of many successful field guides, gardening books, and general interest titles on butterflies, insects, wildflowers, native plants, and wildlife landscaping, including *Butterflies of the Midwest Field Guide; Native Plant Gardening for Birds, Bees & Butter-flies: Upper Midwest; Backyard Bugs; Vibrant Butterflies;* and *Insects & Bugs for Kids*. He lives in Gainesville, Florida, with his wife, Stephanie, and their six cats.

ABOUT ADVENTUREKEEN

We are an independent nature and outdoor activity publisher. Our founding dates back more than 40 years, guided then and now by our love of being in the woods and on the water, by our passion for reading and books, and by the sense of wonder and discovery made possible by spending time recreating outdoors in beautiful places. It is our mission to share that wonder and fun with our readers, especially with those who haven't yet experienced all the physical and mental health benefits that nature and outdoor activity can bring. #bewellbeoutdoors